T0205620

Studies in Computational Intelligence

Volume 790

Series editor

Janusz Kacprzyk, Polish Academy of Sciences, Warsaw, Poland
e-mail: kacprzyk@ibspan.waw.pl

The series "Studies in Computational Intelligence" (SCI) publishes new developments and advances in the various areas of computational intelligence—quickly and with a high quality. The intent is to cover the theory, applications, and design methods of computational intelligence, as embedded in the fields of engineering, computer science, physics and life sciences, as well as the methodologies behind them. The series contains monographs, lecture notes and edited volumes in computational intelligence spanning the areas of neural networks, connectionist systems, genetic algorithms, evolutionary computation, artificial intelligence, cellular automata, self-organizing systems, soft computing, fuzzy systems, and hybrid intelligent systems. Of particular value to both the contributors and the readership are the short publication timeframe and the world-wide distribution, which enable both wide and rapid dissemination of research output.

More information about this series at http://www.springer.com/series/7092

Roger Lee

Editor

Software Engineering, Artificial Intelligence, Networking and Parallel/Distributed Computing

 Springer

Editor
Roger Lee
Software Engineering and Information
  Technology Institute
Central Michigan University
Mount Pleasant, MI, USA

ISSN 1860-949X   ISSN 1860-9503   (electronic)
Studies in Computational Intelligence
ISBN 978-3-030-07488-3   ISBN 978-3-319-98367-7   (eBook)
https://doi.org/10.1007/978-3-319-98367-7

This Springer imprint is published by the registered company Springer Nature Switzerland AG
The registered company address is: Gewerbestrasse 11, 6330 Cham, Switzerland

Foreword

The purpose of the 19th IEEE/ACIS International Conference on Software Engineering, Artificial Intelligence, Networking and Parallel/Distributed Computing (SNPD 2018) on June 27–29, 2018, held at Busan, Korea, is aimed at bringing together researchers and scientists, businessmen and entrepreneurs, teachers, and students to discuss the numerous fields of computer science and to share ideas and information in a meaningful way. This publication captures 13 of the conference's most promising papers, and we impatiently await the important contributions that we know these authors will bring to the field.

In Chapter 1, Donghui Kim, Huimin Mu, and Taesoo Moon examined the relationship between TMS and project team performance by considering technology complexity. Their research found that joint decision making could moderate the relationship between technology complexity and TMS.

In Chapter 2, Yucheng Hou, Sungbae Kang, and Taesoo Moon empirically examined consumers' continuous usage intention of consumer-orientated virtual reality products based on motivation theory. This study proposes several hypotheses that were conducted empirically to test their hypotheses.

In Chapter 3, Woochang Shin proposes a refactoring method for finding duplicate codes used in branch statement and removing them. Based on the proposed method, they also develop and test a prototype tool.

In Chapter 4, Meonghun Lee, Haengkon Kim, and Hyun Yoe designed and implemented a smart farm environment management system based on the ICBM paradigm that can collect and monitor information on crop growth.

In Chapter 5, Hojin Yang and Hyun Yoe examined the transmission of infection of foot-and-mouth disease of livestock that is spread by livestock transportation vehicles. To solve this problem, an I-beacon-based distribution management system was designed. They examined the results of the data collected by this study.

In Chapter 6, Hyeono Choe and Hyun Yoe Hariyama designed the optimal energy management system using ICBM technology to optimize the energy used in cattle shed facilities to grow livestock.

In Chapter 7, Sabrine Malek and Wady Naanaa proposed graph reduction that enables the identification of new polynomially solvable cluster deletion (CD) subproblem. Specifically, they show that if a graph is (butterfly, diamond)-free, then a cluster deletion solution can be found in polynomial time on that graph.

In Chapter 8, Hyungwoo Park, Jong-Bae Kim, Seong-Geon, Bae, Myung-Sook Kim proposed a personal telephone truth discriminator to evaluate whether a telephone voice is a lie. It is obtained by analyzing the relationship between the characteristics of the fundamental frequency (fundamental tone) of the vocal cords and the resonance frequency characteristics of vocal tracks.

In Chapter 9, Lee Ki Bum, Lee Hyung Taek, Kim Jong Yoon, and Gim Gwang Yong examined the factors affecting a user's intention to switch seen in the stage of switching the administrative information system being operated by the local government to a centralized cloud computing, and a research model was identified based on precedent researches in order to achieve the purpose of the study.

In Chapter 10, Liaq Mudassar and Yungcheol Byun proposed customer influx based on parking data as they attempt to predict customer flow of a large grocery store using its parking logs. Their aim is to solve the problem of vehicle parking by predicting the traffic flow within a specific traffic parking lot, with respect to time, using recurrent neural networks (RNNs).

In Chapter 11, Wafa Shafqat and Yungcheol Byun conducted a study that takes a step to identify the hidden themes in crowdfunding site comments to discover the different topics of discussion in scam campaigns, and then, these topics are compared with the topics identified in genuine campaigns.

In Chapter 12, HooKi Lee, HyunHo Jang, SungHwa Han, and GwangYong Gim presented a study that examines the problems of current e-mail security solutions and presents an improved proposed model that is based on security control for spear phishing detection.

In Chapter 13, Junhyun Park, Hwanchul Jung, Jongseok Lee, and Jangwu Jo conducted a survey of state-of-the-art plagiarism detection techniques, called GPLAG. The problem of this technique is that it becomes inefficient when PDG grows. To resolve a problem of time complexity, their paper proposes a way to perform program slicing first and PDG comparison later.

It is our sincere hope that this volume provides stimulation and inspiration and that it will be used as a foundation for works to come.

Haeng-Kon Kim
Huaikou Miao
Takayuki Ito

June 2018

Contents

List of Contributors

Seong-Geon Bae Sori Engineering Lab, Computer Media Information Engineering, Kangnam University, Yongin-si, Gyeonggi-do, Korea

Lee Ki Bum Department of IT Policy and Management, Soongsil University, Seoul, South Korea

Yungcheol Byun Department of Computer Engineering, Jeju National University, Jeju, South Korea

Hyeono Choe Department of Information and Communication Engineering, Sunchon National University, Sunchon, Jeollanam-do, Republic of Korea

GwangYong Gim Department of IT Policy and Management, Soongsil University, Seoul, South Korea

SungHwa Han Department of IT Policy and Management, Soongsil University, Seoul, South Korea

Yucheng Hou Department of International Business, Dongguk University, Gyeongju-si, Gyeongsangbuk-do, Korea

HyunHo Jang Department of IT Policy and Management, Soongsil University, Seoul, South Korea

Jangwu Jo Department of Computer Engineering, Dong-A University, Busan, Korea

Hwanchul Jung VODAS CO., Ltd., Seoul, Korea

Sungbae Kang PARAMITA College, Dongguk University, Gyeongju-si, Gyeongsangbuk-do, Korea

Donghui Kim Department of Techno-Management, Graduate School of Dongguk University, Gyeongju-si, Gyeongsangbuk-do, Korea

Haengkon Kim School of Information Technology, Catholic University of Daegu, Republic of Korea, Gyeongsan, Gyeongbuk-do, Republic of Korea

Jong-Bae Kim Department of Telecommunication Engineering, Soongsil University, Seoul, Republic of Korea

Myung-Sook Kim Department of English Language and Literature, Soongsil University, Seoul, Republic of Korea

HooKi Lee Department of IT Policy and Management, Soongsil University, Seoul, South Korea

Jongseok Lee Department of Computer Engineering, Woosuk University, Wanju, Jeollabuk-do, Korea

Meonghun Lee Department of Agricultural Engineering, National Institute of Agricultural Sciences, Wanju, Jeollabuk-do, Republic of Korea

Sabrine Malek Faculty of Economics and Management of Sfax, Tunis, Tunisia

Taesoo Moon School of Management, Dongguk University, Gyeongju-si, Gyeongsangbuk-do, Korea

Huimin Mu Business School, Guilin University of Technology, Guilin, Guangxi, China

Liaq Mudassar Jeju National University, Jeju, South Korea

Wady Naanaa National Engineering School of Tunis, Tunis, Tunisia

Hyungwoo Park School of Information Technology, Soongsil University, Seoul, Republic of Korea

Junhyun Park Department of Computer Engineering, Dong-A University, Busan, Korea

Wafa Shafqat Department of Computer Engineering, Jeju National University, Jeju, South Korea

Woochang Shin Department of Computer Science, SeoKyeong University, Seoul, Korea

Lee Hyung Taek Department of IT Policy and Management, Soongsil University, Seoul, South Korea

Hojin Yang Department of Information and Communication Engineering, Sunchon National University, Suncheon, Jeollanam-do, Republic of Korea

Hyun Yoe Department of Information and Communication Engineering, Sunchon National University, Suncheon, Jeollanam-do, Republic of Korea

Gim Gwang Yong Department of Business Administration, Soongsil University, Seoul, South Korea

Kim Jong Yoon Department of IT Policy and Management, Soongsil University, Seoul, South Korea

Impact of Transactive Memory Systems on Team Performance and the Moderating Effect of Joint Decision Making

Donghui Kim[1], Huimin Mu[2], and Taesoo Moon[3(✉)]

[1] Department of Techno-Management, Graduate School of Dongguk University,
123, Dongdae-ro, Gyeongju-si, Gyeongsangbuk-do, Korea
kimdh1986@gmail.com
[2] Business School, Guilin University of Technology,
Jian'gan Road #12, Guilin, Guangxi, China
muhuimin_222@126.com
[3] School of Management, Dongguk University,
Dongdaero 123, Gyeongju-si, Gyeongsangbuk-do, Korea
tsmoon@dongguk.ac.kr

Abstract. Many studies related to transactive memory systems (TMS) have examined project team efficiency. Specially, information system (IS) development projects require high efficiency within limited budgets and time schedules, even though technology applications cannot easily resolve complex problems. Our study examines the relationship between TMS and project team performance by considering technology complexity. In order to identify the relationship between variables, this study conducted a survey involving 83 respondents. Empirical results showed that technology complexity had a positive impact on TMS, and TMS had a positive impact on project team performance. Our research found that joint decision making could moderate the relationship between technology complexity and TMS.

Keywords: Transactive memory systems
Information system development project · Technology complexity
Joint decision making · Project team performance

1 Introduction

Modern companies are increasingly aware of the significance of information and knowledge, and effectively utilize these resources for achieving business purposes. They also realize that strategic use of information systems is a critical factor for the success of business collaboration activities. The system integration (SI) industry has rapidly grown along with the trend of IT diffusion. However, most SI firms have experienced the lack of IT expertise for new IS projects and the lack of IT competence in new market development.

Information systems (IS) development projects are very difficult to succeed because user companies have their unique and unstructured requests for implementing their information systems. For maintaining SI business, they need to search for methods to

© Springer Nature Switzerland AG 2019
R. Lee (Ed.): SNPD 2018, SCI 790, pp. 1–13, 2019.
https://doi.org/10.1007/978-3-319-98367-7_1

increase teamwork quality and efficiency. Transactive memory systems (TMS) were adopted by several studies for enhancing project and team performance in hospitals [21], financial companies, and telecom companies [8]. These studies examined the impacts of TMS on team characteristics such as task interdependence [11] and group training [22]. However, there have been few studies related to IS development project and few studies have considered team environment.

IS development projects need high efficiency within limited budget and time schedule, even though technology applications cannot easily resolve various complex problems in the organization. To assure success of IS development projects within budget and desired timeframe, effective and efficient teamwork is required. However, the success of IS development projects is very difficult because the projects often experience issues integrating various technologies and client systems. This phenomenon increases the technical complexity of projects.

This study aims to suggest a research framework to increase efficiency and effectiveness of IS development project teams. To achieve this goal, characteristics of technology complexity and joint decision making as project teams are drawn from previous studies. TMSs are adopted to increase teamwork quality in IS development projects, and to identify its relationship with project team performance.

2 Literature Review and Hypothesis Development

2.1 Transactive Memory Systems

TMS describes the active use of transactive memory by more than two people to cooperatively store, retrieve, and communicate information [16]. Wegner [30] first introduced the concept of transactive memory to explain the behavior of couples. Wegner and his colleague suggested that partners in intimate relationships cultivate one another as external memory aids and develop a "shared system for encoding, storing, and retrieving information" [31].

Based on Wegner's studies, Lewis [17] developed the measurement of TMS. The author argued that the TMS construct is a second-order factor consisting of specialization, credibility and coordination. IS development projects feature natural complexity because it encompasses both organizational issues and technological factors. We suggest a research model regarding IS development projects based on an extensive literature review. Figure 1 shows the research model in this study.

2.2 Technology Complexity and TMS

IS development projects involve activities related to the IS implementation process including user requirements, system design, coding, programming, testing, and maintenance [14]. Due to IS development projects containing complex activities, IS development projects feature complexity [1, 36]. Thus, complexity is an important factor which influences the success of a project and is considered a key factor in IT project management [35].

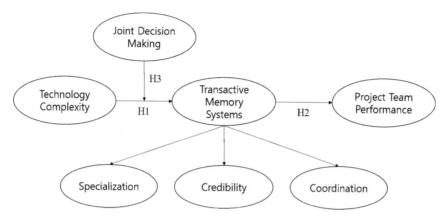

Fig. 1. Research model

Baccarini [1] argued that project complexity has two dimensions: organizational complexity and technological complexity. Organizational complexity results from the organizational hierarchical structure, organizational units, and task structure. Technological complexity is affected by the number of tasks for the project outcome, required number of specialized areas, and interdependent features.

IS development projects require a variety of programing languages for legacy applications and systems on different operating systems, and complicated web language for user interface. For the execution of IS development projects, bigger projects need a greater number of necessary tasks, expertise and technologies. Increasing number of tasks, expertise and technologies are further challenges in successful outcomes. That makes IS development project more complicated.

TMS acts as a collective filter to enable organizational groups to mitigate the potential for information overload that occurs by specialized technology and knowledge [34]. TMS plays a role of perceived division of labor in the specialized field and can be seen as a mechanism of information collection and management. Therefore, members of projects that have complexity tend to communicate with others to search for information and knowledge. Communication within project team members facilitates the development of TMS [12] Based on the literature review, the following hypothesis is proposed:

H1: Technology complexity of IS development projects has a positive impact on transactive memory systems.

2.3 TMS and Team Performance

TMS impacts project team performance by enhancing information acquisition, learning and communication among team members [17]. Project team performance includes accomplishing user requirements within the project budget and time period, and measured by quality of outcome, efficiency of project management, and timeliness. [5, 13, 19].

TMS has significant influence on project and team performance in various industries. Choi et al. [5] examined the effect of transactive memory systems on team performance in oil companies and steel companies. Wang et al. [29] examined TMS impact on team performance through knowledge transfer in knowledge- intensive industries in China such as manufacturing, IT, education, and construction. Chiang et al. [3] argued TMS can improve new product development performance. Chiang and his colleagues found that TMS can improve the performance of new products developed by the NPD team in Taiwanese electronics manufacturers. In addition, impact of TMS on team and project performance have been verified in various industries, including automobile parts manufacturing [28], online contents [18], IT consulting and system integration [15].

Hsu et al. [13] suggested that TMS in IS development projects could influence communication and task coordination, as well as team performance. TMS can promote project team performance by reducing the time to search for and acquire information needed for completion of tasks, thereby improving the efficiency and effectiveness of the tasks. Based on the literature review and reasoning, the following hypothesis is proposed:

H2: Transactive memory systems of IS development teams have a positive impact on project team performance.

2.4 Moderating Effect of Joint Decision Making

Many companies introduce partnership sourcing for effective timeliness and responsiveness, which may improve results [25]. Partnership refers to the degree to which the dyad party jointly takes responsibility for solving problems [2]. Previous studies have argued that cognition of belonging to the same team or organization is an important drive factor of team performance [6].

Joint decision making, referring to the degree to which team members jointly make decisions about key issues or challenges [19], is evidence of partnerships and could enhance the degree of participative management [9]. In the IS development project case, joint decision making provides equal opportunities that allow team members to express their ideas. In addition, team members participate in the decision making process which can reduce misunderstanding or conflicts. These advantages shorten the time and budget required for tasks and projects, and contribute to quality improvements.

Project team members participating in joint decision making could facilitate the development of knowledge, which results in enhanced communication among team members. Thus, their awareness of the team increases [27]. This transactive partnership and joint decision making requires communication with team members for enhanced decision making, thinking and behavior. Through the joint decision making process, individuals of project teams could enhance understanding of other team members and develop transactive memory which is an essential part of TMS. Based on the literature review, we hypothesize that joint decision making has a moderating effect on the relationship between technology complexity and TMS.

H3: Joint decision making within IS development project teams will moderate the relationship between technology complexity and TMS

3 Research Methodology

3.1 Sampling and Measurement Model

Measurement items of each construct were extracted from prior studies. In detail, items of technology complexity were adopted from Xia and Lee [36], TMS from Lewis [16], project team performance from Choi et al. [5], Hsu et al. [13], and Lin et al. [19], and joint decision making from Subramani and Venkatraman [26] and Lin et al. [19]. All measurement items were measured by five-point Likert-type scales ranging from strongly disagree (1) to strongly agree (5). Data from 104 respondents was collected from IS development project team members in IT companies, with 83 valid answers gathered and analyzed in this study. Table 1 shows the demographic characteristics of respondents.

Table 1. Demographic characteristics

Characteristics		Frequency	Percent
Industry[a]	SW development	44	53.0
	System integration	42	50.6
	Consulting	27	32.5
	HW facility	9	10.8
	Information security	13	15.7
	Network	8	9.6
	Database	19	22.9
	Etc.	10	12.0
Department/team	Information strategy plan	7	8.4
	SW development	29	34.9
	Database	4	4.8
	Network	2	2.4
	HW facility	4	4.8
	Sales	4	4.8
	Consulting	19	22.9
	Project management	5	6.0
	Quality control	7	8.4
	Etc.	2	2.4
Project team size	1–6	17	20.5
	6–10	38	45.8
	11–20	20	24.1
	21–50	5	6.0
	50 over	3	3.6
Total		83	100

[a]Multiple choice

The question concerning the industry that respondents belong to was measured using a multiple choice format. 44 respondents belonged to SW development, and 42 respondents belonged to system integration. In addition, 27 respondents belonged to consulting, 19 to database, and 13 belonged to information security. The question related to department found the system development department had the highest number of respondents (22) and the consulting department had 19 respondents.

Although we used a relatively small sample for analysis, the partial least squares (PLS) method can model latent constructs with small- to medium-sized data sets that do not necessarily follow a normal distribution [4, 7]. Generally, in social science studies, the cutoff value of cross loading value is 0.6, AVE is 0.5, Cronbach's alpha is 0.7, and composite reliability is 0.7.

The Cronbach's alpha of technology complexity is 0.711. All research variables exceeded the threshold of 0.7 for Cronbach's alpha (specialization = 0.770, credibility = 0.805, coordination = 0.784, joint decision making = 0.766, project team performance = 0.830).

The results of validity and reliability test, Cronbach's alpha (from 0.711 to 0.810), composite reliability (from 0.822 to 0.878), and AVE (from 0.539 to 0.707) indicate sufficient convergent validity for all constructs. Table 2 and 3 show validity and reliability of research variables.

Furthermore, analysis results on the discriminant validity in this study are shown in Table 3. The AVE square root of each variable showed a lower correlation coefficient than the other variables. Therefore, there is no problem with the discriminant validity.

TMS is a second order factor encompassing 3 dimensions; that is, specialization, credibility, and coordination [16]. We tested the validity of the second order factor following the method in Wetzels et al. [33]. Wetzels and his colleagues suggested a second order model is valid when cross loading values of second order factors (in our model, specialization, credibility and coordination) are larger than that of the first order factor. Figure 2 and Table 4 shows the cross loading values of measurement items for TMS, specialization, credibility and coordination.

3.2 Hypothesis Testing

We examined the significance of the paths in the structural model. Analysis results for the structural model are presented in Table 5. According to the results, technology complexity has a positive impact on TMS. The path coefficient is 0.775 and t-value is 14.739, so H1 is accepted. H2 is also accepted in that TMS has a significant positive impact on project team performance, with the path coefficient of 0.483 and t-value of 6.072. In Model 0, R^2 of TMS is 0.603 and R^2 of project team performance is 0.234. This R^2 shows that technology is one of the most impact factor of TMS.

For testing the moderating effect of joint decision making, this study used the F^2 test. With the moderating variable (Model 2), explaining power of TMS was increased (R^2 difference is 0.097). According to the results (F^2 is 0.323), joint decision making has a moderating effect between technology complexity and TMS.

Table 2. Validity and reliability

Variable	Measurement item	Cross loading	Cronbach's alpha	Composite reliability	AVE
Technology complexity	The project involved multiple software tools	0.825	0.711	0.822	0.539
	The project involved multiple OS and system development environments	0.685			
	The project team was cross-functional	0.638			
	The project involved integration between legacy systems or applications	0.773			
Specialization	Each team member has specialized knowledge of some aspect of our project	0.811	0.770	0.853	0.594
	Each team member has knowledge about an aspect of the project that no other team member has	0.781			
	Different team members are responsible for expertise in different areas	0.790			
	Each team member know which team members have expertise in specific areas	0.695			
Credibility	Each team member trusts the other members' expertise and know-how	0.750	0.805	0.865	0.563
	Each team member trusts the other members' knowledge about the project was credible	0.844			
	Each team member was confident relying on the information that	0.698			

(*continued*)

Table 2. (*continued*)

Variable	Measurement item	Cross loading	Cronbach's alpha	Composite reliability	AVE
	other team members brought to the discussion				
	When other members gave information, each team member trusts that information without double-check	0.763			
Coordination	Our team had very few misunderstandings about what to do	0.646	0.784	0.860	0.609
	Each team member doesn't need to backtrack and start over	0.780			
	Our team accomplished the task smoothly and efficiently	0.818			
	Each team member carries out task without conflict	0.860			
Joint decision making	Our team members developed task strategies together	0.818	0.766	0.851	0.590
	Our team members set task goals together	0.802			
	Our team members diagnosed problems together	0.770			
	Our team members collected required data together	0.673			
Project team performance	Our team has successfully completed the project to meet its goals	0.811	0.830	0.880	0.596
	The results of the our project team developed can be evaluated as excellent	0.695			
	Our team adherence to project schedule	0.699			

(*continued*)

Table 2. (*continued*)

Variable	Measurement item	Cross loading	Cronbach's alpha	Composite reliability	AVE
	Our team adherence to each project work plan	0.822			
	Our team accomplished project efficiently	0.823			

Table 3. Discriminant validity

	Mean	S.D.	TC	Sp	Cr	Co	Jo	PTP
TC	3.792	0.562	**0.734**					
Sp	3.921	0.555	0.664	**0.770**				
Cr	3.819	0.580	0.635	0.576	**0.797**			
Co	3.443	0.765	0.536	0.390	0.376	**0.780**		
JD	3.804	0.584	0.426	0.486	0.480	0.431	**0.768**	
PTP	3.858	0.619	0.305	0.337	0.321	0.508	0.409	**0.772**

Note: TC: Technology Complexity, TMS: Transactive Memory Systems, JDM: Joint Decision Making, PTP: Project Team Performance

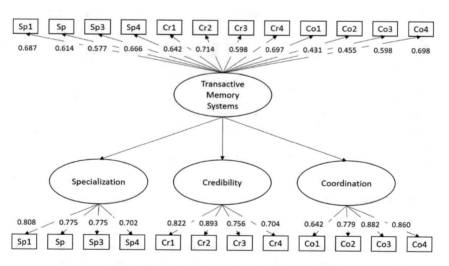

Fig. 2. Validity test of second order factor model

Table 4. Comparing cross loadings between 1st order factor model and 2nd order factor model

Measurements	Second order factor model		First order factor model	
	Variable	Cross loading	Variable	Cross loading
Sp1	Specialization	0.808	Transactive memory systems	0.687
Sp2		0.775		0.614
SP3		0.792		0.577
SP4		0.702		0.666
Cr1	Credibility	0.822		0.642
Cr2		0.893		0.714
Cr3		0.756		0.598
Cr4		0.704		0.697
Co1	Coordination	0.642		0.431
Co2		0.779		0.455
Co3		0.882		0.598
Co4		0.860		0.698

Table 5. Result of hypotheses test

Hypothesis		Model 0	Model 1	Model 2
H1	TC → TMS	0.775 (14.739)**	0.642(8.142)**	0.628(8.024)**
H2	TMS → PTP	0.483(6.072)**	0.483(5.982)**	0.483(5.924)**
Moderate variable	JDM → TMS	-	0.314(3.391)**	0.287(3.323)**
H3	TC * JDM → TMS	-	-	0.116(1.965)*
		TMS | PTP	TMS	TMS
R^2		0.603 | 0.234	0.680	0.700
R^2 Difference			0.077	0.097
F^2				0.323
F^2 Value				31.687

Note. TC: Technology Complexity, TMS: Transactive Memory Systems, JDM: Joint Decision Making, PTP: Project Team Performance, *P < 0.05, **P < 0.01, ( ): t-value, F (0.05,1,83) = 3.956, F(0.01,1,83) = 6.95,
Effect size. small: < 0.02, medium: < 0.15, Large: < 0.35

4   Discussion and Conclusion

This research examined how to increase IS development project team performance considering the complex technological nature of IS development projects. We collected 83 surveys from Korean IT firms, and verified the research hypothesis through empirical analysis. Results show complex technology positively impacts TMS, and TMS enhances project team performance. In addition, joint decision making moderates technology complexity and TMS. Our findings have academic and practical implications.

First, technology complexity has a positive influence on TMS. Technology complexity has generally been regarded as project management and has been recognized as a factor of project complexity. The complexity of IS development is manifested by the historically high failure rate of IS development projects [36]. However, results of our study show that technology complexity positively influences the development of TMS. The positive impact of technology complexity is a valuable result in project management and management information systems areas. This evidence shows that technology complexity can be mitigated through TMS.

Second, TMS has a positive influence on project team performance. In this paper, project team performance was measured using operating performance and output quality in a non-financial performance perspective. TMS can achieve superior outcomes within limited budgets, time and project scope via increasing project efficiency. This result is similar with previous studies. Project managers promote members' transactive memory at the organization or group level.

Third, joint decision making moderates the relationship between technology complexity and TMS. Joint decision making was mainly considered a mediating variable in previous researches in this area. However, our finding shows that project managers should encourage team members to participate in decision making.

This research explores methods of increasing IS development project team performance. To achieve this goal, technology complexity, and joint decision making are drawn from previous studies as project team environment and characteristics. TMSs are adopted to increase teamwork quality in IS development projects, and to identify its relationship with project team performance. A total of 104 respondents participated in our survey, among which 83 completed surveys were analyzed using Smart PLS.

Technology complexity, adopted from prior literature, positively influences TMS. Furthermore, TMS exerts a significant positive impact on project team performance. TMS is a second order factor which encompasses three dimensions – specialization, credibility, and coordination. Differing from previous studies that considered joint decision making as a mediating variable, our study provides empirical evidence that joint decision making moderates the relationship between technology complexity and transactive memory systems.

References

1. Baccarini, D.: The concept of project complexity – a review. Int. J. Proj. Manag. **14**(4), 201–204 (1996)
2. Biehl, M., Kim, H., Wade, M.: Relationships among the academic business disciplines: a multi-method citation analysis. Omega **34**(4), 359–371 (2006)
3. Chaing, Y.H., Shih, H.A., Hsu, C.C.: High commitment work system, transactive memory system, and new product performance. J. Bus. Res. **67**, 631–640 (2014)
4. Chin, W.W.: The partial least squares approach to structural equation modeling. In: Marcoulides, G.A. (ed.) Modern Methods for Business Research, pp. 295–336. Lawrence Erlbaum, London (1998)

5. Choi, S.Y., Lee, H., Yoo, Y.: The impact of information technology and transactive memory systems on knowledge sharing, application, and team performance: a field study. MIS Q. **34**(4), 855–870 (2010)
6. DeChurch, L.A., Mesmer-Magnus, J.R.: The cognitive underpinnings of effective teamwork: a meta-analysis. J. Appl. Psychol. **95**(1), 32–53 (2010)
7. Fornell, C., Larcker, D.F.: Evaluating structural equation models with unobservable variables and measurement error. J. Mark. Res. **28**, 39–50 (1981)
8. Hammedi, W., Riel, A.C.R.V., Sasovova, Z.: Improving screening decision making through transactive memory systems: a field study. J. Prod. Innov. Manag. **30**(2), 316–330 (2013)
9. Heide, J.B., John, G.: Alliances in industrial purchasing: the determinants of joint action in buyer-supplier relationships. J. Mark. Res. **27**(1), 24–36 (1990)
10. Hoegl, M., Gemuenden, H.G.: Teamwork quality and the success of innovative projects: a theoretical concept and empirical evidence. Organ. Sci. **12**(4), 435–449 (2001)
11. Hollingshead, A.B.: Cognitive interdependence and convergent expectations in transactive memory. J. Pers. Soc. Psychol. **81**(6), 1080–1089 (2001)
12. Hollingshead, A.B., Brandon, D.P.: Potential benefits of communication in transactive memory systems. Hum. Commun. Res. **29**(4), 607–615 (2003)
13. Hsu, J.S.C., Shih, S.P., Chiang, J.C., Liu, U.Y.C.: The impact of transactive memory systems on IS development teams' coordination, communication, and performance. Int. J. Proj. Manag. **30**(3), 329–340 (2012)
14. Jang, S.B., Kwahk, K.Y.: The effects of IT project risk management factors on project performance. Korean Manag. Sci. Rev. **28**(2), 31–51 (2011)
15. Kim, D.H., Kang, S.B., Moon, T.S.: The impact of transactive memory systems and expertise integration on project team performance: focused on information system development. J. Internet Electron. Commer. Res. **15**(4), 205–222 (2015)
16. Lewis, K.: Measuring transactive memory systems in the field:scale development and validation. J. Appl. Psychol. **88**(4), 587–603 (2003)
17. Lewis, K., Lange, D., Gillis, L.: Transactive memory systems, learning and learning transfer. Organ. Sci. **16**(6), 581–598 (2005)
18. Lim, H.J., Kang, H.R.: Product development teams effectiveness: the role of transactive memory system. J. Organ. Manag. **30**(1), 31–58 (2006)
19. Lin, T.C., Hsu, J.S.C., Cheng, K.T., Wu, S.: Understanding the role of behavioural integration in ISD teams: an extension of transactive memory systems concept. Inf. Syst. J. **22**(3), 211–234 (2012)
20. McKeen, J.D., Guimaraes, T., Wetherbe, J.C.: The relationship between user participation and user satisfaction: an investigation of four contingency factors. MIS Q. **18**(4), 427–451 (1994)
21. Michinov, E., Olivier-Chiron, E., Rusch, E., Chiron, B.: Influence of Transactive memory on perived performance jobsatisfacion and identification in anaesthesia teams. Br. J. Anaesth. **100**(3), 327–332 (2008)
22. Molreland, R.L., Argote, L., Krishnan, R.: Socially shared cognition at work: transactive memory and group performance. In: Nye, J.L., Brower, A.M. (eds.) What's Social about Social Cognition? Research on Socially Shared Cognition in Small Groups. Sage, Thousand Oaks (1996)
23. Ribbers, M.A.P., Schoo, K.C.: Program management and complexity of ERP implementations. Eng. Manag. J. **14**(2), 45–52 (2002)
24. Simsek, Z., Lubatkin, M.L., Dino, R.N.: Modeling the multilevel determinants of top management team behavioral integration. Acad. Manag. J. **48**(1), 69–84 (2005)
25. Stanley, L.L., Wisner, J.D.: Service quality along the supply chain: implications for purchasing. J. Oper. Manag. **19**(3), 287–306 (2001)

26. Subramani, M.R., Venkatraman, N.: Safeguarding investments in asymmetric interorganizational relationships: theory and evidence. Acad. Manag. J. **46**(1), 46–62 (2003)
27. Van De Ven, A.H., Delbecq, A.L., Koenig, R.J.: Determinants of coordination modes within organizations. Am. Sociol. Rev. **41**(2), 322–338 (1976)
28. Wang, F.Z., Choi, S.M., Moon, T.S.: A study on the influence of Transactive Memory Systems (TMS) and knowledge use on team performance: focus on automobile parts industry. J. Internet Electron. Commer. Res. **14**(1), 41–63 (2014)
29. Wang, Y., Huang, Q., Davison, R.M., Yang, F.: Effect of transactive memory systems on team performance mediated by knowledge transfer. Int. J. Inf. Manag. **41**, 65–79 (2018)
30. Wegner, D.M.: Transactive memory: a contemporary analysis of the group mind. In: Mullen, B., Goethals, G.R. (eds.) Theories of Group Behavior, pp. 185–208. Springer, New York (1987)
31. Wegner, M.D., Erber, R., Raymond, P.: Transactive memory in close relationship. J. Pers. Soc. Psychol. **61**(6), 923–929 (1991)
32. Weick, K.E., Roberts, K.H.: Collective mind in organizations: heedful interrelating on flight decks. Adm. Sci. Q. **38**(3), 357–381 (1993)
33. Wetzels, M., Odekerken-Schröder, G., Van Oppen, C.: Using PLS path modeling for assessing hierarchical construct models: guidelines and empirical illustration. MIS Q. **33**(1), 177–195 (2009)
34. Whelan, E., Teigland, R.: Transactive memory systems as a collective filter for mitigating information overload in digitally enabled organizational groups. Inf. Organ. **23**(3), 177–197 (2013)
35. Whitney, K.M., Daniels, C.B.: The root cause of failure in complex IT projects: complexity itself. Procedia Comput. Sci. **20**, 325–330 (2013)
36. Xia, W., Lee, G.: Complexity of information systems development projects: conceptualization and measurement development. J. Manag. Inf. Syst. **22**(1), 45–83 (2005)

Customers' Continuous Usage Intention of Virtual Reality (VR) Product

Yucheng Hou[1(✉)], Sungbae Kang[2], and Taesoo Moon[3]

[1] Department of International Business, Dongguk University,
123, Dongdae-ro, Gyeongju-si, Gyeongsangbuk-do, Korea
yc-hou@foxmail.com
[2] PARAMITA College, Dongguk University,
123, Dongdae-ro, Gyeongju-si, Gyeongsangbuk-do, Korea
sbkang@dongguk.ac.kr
[3] School of Management, Dongguk University,
123 Dongdae-ro, Gyeongju-si, Gyeongsangbuk-do, Korea
tsmoon@dongguk.ac.kr

Abstract. A new wave of virtual reality (VR) developments promise to make the technology mainstream. Virtual reality captures nearly every sensation that a person experiences, not only as an audience but as a participant. Mass adoption of virtual reality is applying to existing industries. When the first generation consumer-oriented virtual reality products were introduced, some industries injected virtual reality products into their business, such as tourism, movie, gaming, and shopping. Hence, understanding consumers' continuous usage intention of virtual reality product is important. Due to people are willing to use virtual reality products in their daily life, the real virtual reality era will eventually be realized. However, consumers' continuous usage intention of virtual reality products is currently not well understood. Prior studies have predominantly focused on the realization of virtual reality technology and its application in specific areas. There are limited studies aimed at consumers' acceptance and continuance usage intention of virtual reality products. To complement this gap, the purpose of this study is to empirically examine consumers' continuous usage intention of consumer-oriented virtual reality products based on motivation theory. This study proposes several hypotheses and conducted empirically tests our hypotheses.

Keywords: Virtual reality · Continuous usage intention
Virtual reality product · Motivation theory

1   Introduction

On March 26, 2014, Facebook announced that they would spend 2 billion dollars in the acquisition of the immersive virtual reality technology company, Oculus VR. This event brought virtual reality into people's consciousness again. They were amazed at what kind of product enticed Facebook to spend 2 billion dollars in its acquisition. Over the next 2 years, Samsung and HTC quickly launched their virtual reality products. Due to virtual reality captures nearly every single sensation that a person

© Springer Nature Switzerland AG 2019
R. Lee (Ed.): SNPD 2018, SCI 790, pp. 14–26, 2019.
https://doi.org/10.1007/978-3-319-98367-7_2

experiences not only as an audience but as a participant, many researchers regard virtual reality as the 'ultimate' media. Furthermore, the mass adoption of virtual reality products is applying to existing industries. For example, Whyte, Bouchlaghem, Thorpe, and McCaffer [27] stated that virtual reality has the potential to improve the visualization of building design and construction. Huang, Rauch, and Liaw [14] argued that virtual reality can be used for education but needs sufficient contents.

In fact, after Facebook merged with Oculus and the first-generation consumer-oriented virtual reality products had been introduced, some industry leaders tried to inject virtual reality elements into their business. In the second half of 2015, Universal Studios and Disney announced that they entered the field of virtual reality film. Furthermore, in November 2015, the New York Times published "NYT VR", which utilizes virtual reality in the media industry. In March 2016, Alibaba announced that Buy + plan had been launched which incorporated virtual reality in e-shopping. In China, virtual reality will also be used in live events. For example, in January 2017, CCTV (China Central Television) added virtual reality elements in the Spring Festival Gala.

As many industries want to inject virtual reality into their industry chain, understanding consumers' continuous usage intention of first-generation virtual reality products is important. Due to people are willing to use virtual reality products in their daily life, the real virtual reality era will eventually be realized. However, prior studies have predominantly focused on the realization of virtual reality technology [4, 5, 7] and its application in specific areas [11, 14, 27]. No prior studies have examined the consumers' continuous usage intention of consumer-oriented virtual reality products. Hence, this study aims to complement this gap. This study focuses on consumers' continuous usage intention of consumer-oriented virtual reality products. Furthermore, we try to reveal why people are willing to use virtual reality products, what factors influence the continuous usage intention, and how these key variables differ the process of consumer decision making. This study suggests a research model to identify the relationship between influence factors, customer motivation and continuous usage intention.

2 Literature Review

2.1 Virtual Reality

According to the Oxford English Dictionary, virtual reality refers to "The computer-generated simulation of a three dimensional image or environment that can be interacted with in a seemingly real or physical way by a person using special electronic equipment, such as a helmet with a screen inside or gloves fitted with sensors." Figueiredo, BÖhm, and Teixeira [10] defined virtual reality as a computer generated environment that gives the illusion of being immersed in a real system. Bryson [5] provided another definition about virtual reality, which is the use of computers and human-computer interfaces to create the effect of a three dimensional world containing interactive objects with a strong sense of three dimensional presence. Differing from the definition in 1993 [5], Bryson [5] emphasized interactive objects should be added into virtual reality. Hence, in our opinion, the characteristics of VR not only includes immersion, but also interaction. Immersion describes the degree of simulation provided to users while defining the depth of the user can experience the reality. As for interaction, it permits users to not only receive information from the VR system, but also create, control, observe and communicate with the system.

However, the target population of our study is consumers, and consumer immersion and interaction is not easy to understand. Hence, in this study, we accept the definition of virtual reality that borrows from Steuer (1992), who was the first to define virtual reality according to the concept of telepresence. Telepresence is the experience of presence in an environment by means of a communication medium [24]. Simply, telepresence means "be there". Users can feel they are really in the environment generated by virtual reality products. According to Steuer [24], immersion means the depth of telepresence, and interactivity is one of the characteristic of telepresence. Hence, we consider telepresence as the characteristic of virtual reality products in this study.

Virtual reality was originally developed to design games for military purposes, but is currently used in a variety of fields [1]. Huang, Rauch, and Liaw [14] found immersion had greater contribution than interaction in motivation to use virtual reality products in the educational domain. Ahn, Cho, and Jeong [1] found VR can be used in the travel industry. Bertrand and Bouchard [3] studied people who are favorable to virtual reality use, and found perceived usefulness is the only significant predictor of usage intention. Furthermore, Jung et al. [15] investigated antecedents and consequences of consumers' virtual reality roller coaster usage, which is a popular experience content of consumer-oriented virtual reality products. They found entertainment value is important.

2.2 Motivation Theory

Prior studies have widely used motivation theory to explain an individual's behavior in accepting information technology. Motivations are often divided into intrinsic and extrinsic motivation [8, 22]. Ryan and Deci [22] defined intrinsic motivation as doing something because it is inherently enjoyable or interesting. Davis, Bagozzi, and Warshaw [8] defined extrinsic motivation as the performance of an activity because it is perceived to be instrumental in achieving valued outcomes that are distinct from the activity itself.

Prior studies have exclusively used perceived enjoyment to represent intrinsic motivation and perceived usefulness to represent extrinsic motivation in discussions of how motivators influence individuals' information technology acceptance behavior [8, 16, 18, 23, 25, 26]. Van der Heijden [25] found intrinsic motivation (perceived enjoyment) is a stronger determinant of intention to use than extrinsic motivation in the hedonic domain. Lin and Lu [18] stated that the main use of social networking sites is in the hedonic domain and found perceived enjoyment is the most influential factor in people's continued use of social networking sites. Also Seol, Lee, Yu, and Zo [23] and Wamba, Bhattacharya, Trinchera, and Ngai [26] found for social media, which is in the hedonic domain, intrinsic motivation is a stronger determinant for usage intention than extrinsic motivation. Furthermore, Kim, Hwang, Zo, and Lee [16] also used motivation theory to investigate users' continuance intention towards smart phone augmented reality.

From prior studies, we found that motivation theory can be used to explain hedonic technology and can also be used to research new information technology continuous usage intention. Nowadays, virtual reality products are a new technology product, and the main use in consumers' daily life is playing games and watching movies. Consumers of virtual reality products prefer entertainment products. Hence, this study considers motivation theory as the main theory.

3 Research Model and Hypotheses

3.1 Research Model

The main structure of this study is based on motivation theory. Prior studies found that motivation theory can be used to explain hedonic technologies and can be effectively used to examine new information technology continuous usage intention.

Nowadays, consumer-oriented virtual reality products are more inclined to be entertainment-based products. Hence, enjoyment is important. Virtual reality products as new technology products have some useful areas. Hence, usefulness is also important for consumers in using virtual reality products. In this study, we consider perceived enjoyment as intrinsic motivation and perceived usefulness as extrinsic motivation.

Virtual reality products need content to support its use. Sufficient high quality content can make experience enjoyment and perceive usefulness. Telepresence is one of the most significant characteristic of virtual reality. Higher telepresence can give users a more real virtual experience, and more real virtual experiences give users more enjoyment and usefulness. Thus, we consider virtual reality content quality and telepresence as independent variables in this study. Our research model is shown in Fig. 1.

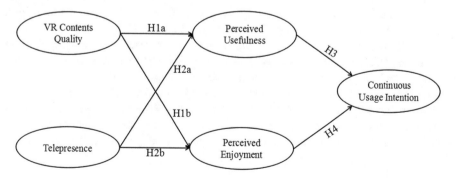

Fig. 1. Research model

3.2   Research Hypotheses

According to media richness theory, the quality of information exchanged across a medium is critical. Virtual reality products need high quality content to support its use. Content quality is the degree of fitness of content for users in terms of adequacy, currentness, and reliability [2]. Low quality content is distracting because it will increases users' search and content processing costs. Users will waste time and effort on useless and boring content. In fact, consumers will be motivated to use virtual reality products when a virtual reality product provides high quality contents.

In Yu, Lee, Ha, and Zo [28] study on media tablets, they found content quality has a positive influence on both perceived usefulness and perceived enjoyment. Almaiah, Jalil, and Man [2] also found content quality is a strong predictor of perceived usefulness. In the context of continuance of using SNS, Seol, Lee, Yu, and Zo [23] also stated content quality has a positive influence on both perceived usefulness and perceived enjoyment. In Kim, Hwang, Zo, and Lee [16] 's study on smart phone augmented reality applications the same results were obtained. On the basis of these results, we expect that for new technology products, content quality is important, and high quality content will provide consumers with a high perception of usefulness and enjoyment. Hence, we proposed the following hypotheses:

H1a: *The content quality of virtual reality products has a positive influence on perceived usefulness.*

H1b: *The content quality of virtual reality products has a positive influence on perceived enjoyment*

According to Draper, Kaber, and Usher [9], telepresence includes two definitions in common use: the cybernetic, and the experiential. In the cybernetic definition, telepresence is an index of quality of the human-machine interface. In the experiential definition, telepresence is a mental state in which a user feels physically present within the computer mediated environment. Hence, in our opinion, telepresence encompasses immersion and interaction. When consumers use virtual reality products, whether the product can make them feel "being there" is important for their satisfaction.

Klein [17] and Park, Hyun, Fairhurst, and Lee [20] found that the higher the level of telepresence, the more real the virtual experience feels to users. And undoubtedly, the more real the virtual experience feels results in a higher level of motivation to use it. In Klein [17] 's study, people with a higher level of telepresence were more likely to find the experience useful than those with a lower level of telepresence. Park, Hyun, Fairhurst, and Lee [20] also found telepresence has a positive influence on perceived usefulness. In the hedonic domain, Pelet, Ettis, and Cowart [21] found telepresence was an important predictor of perceived enjoyment. In Kim, Hwang, Zo, and Lee [16], they used interactivity and vividness as two sub-factors of telepresence and found visual quality positively effects both perceived enjoyment and perceived usefulness, and interactivity positively affected perceived enjoyment. Based on previous studies, we propose the following hypotheses:

H2a: *The telepresence of virtual reality products has a positive influence on perceived usefulness.*

H2b: *The telepresence of virtual reality products has a positive influence on perceived enjoyment*

Davis, Bagozzi, and Warshaw [8] indicated perceived usefulness is the strongest determinant of the adoption of systems or new technology. In the virtual reality domain, perceived usefulness has also been consider in prior researches [3]. Although in these studies the dependent variable is usage intention, we believe in the virtual reality domain it is plausible that perceived usefulness also positively influences continuous usage intention.

In the literature concerning what motivates people to continue to use smart phone augmented reality applications, Kim, Hwang, Zo, and Lee [16] stated that perceived usefulness has a strongly positive influence on continuous usage intention. In Lin and Lu [18], perceived usefulness had a positive effect on continuous usage intention. Based on prior studies, perceived usefulness is a strong motivation of the continuous usage intention of new technology products. Hence, we hypothesize that:

H3: *Perceived usefulness of virtual reality products has a positive influence on continuous usage intention*

Van der Heijden [25] found perceived enjoyment is the main motivation for hedonic information system use. In the virtual reality domain, Jung et al. [15] found entertainment value is important. In the online gaming context, Merikivi, Tuunainen, and Nguyen [19] stated that an important motivation is to obtain pleasure or enjoyment, and the greater enjoyment users experience results in increased motivation to play. Nowadays, the main use of virtual reality products for consumers is playing games and watching movies. Hence, in our opinion, intrinsic motivation has a positive influence on continuous usage intention.

Lin and Lu [18] argued that perceived enjoyment is an important variable in hedonic orientation and positively influences continuous usage intention. Seol, Lee, Yu, and Zo [23] found when individuals experience enjoyment while using social networking services, they are more likely to use them. In fact, when consumers use virtual reality products, enjoyable and fun experiences evoke positive feelings that generate a high degree of continuous usage intention. Hence, perceived enjoyment is an important motivator of continuous usage intention of virtual reality products, and the following hypothesis is proposed:

H4: *Perceived enjoyment of virtual reality products has a positive influence on continuous usage intention*

4   Research Design and Analysis

4.1   Operational Definition and Measurement

The measures used to operationalize the constructs included in the investigated models are mainly adapted from previous studies with modifications to fit the target contexts. The measurements of each construct is shown in Table 1.

Table 1. Operational definition and measurement

Variable	Measurements	Reference
VR content quality	The VR contents such as games, movies, and travel are various	[2, 28]
	The VR contents compared to other devices have high quality.	
	The contents of VR product are accurate and objective.	
	The VR applications that I can choose are various.	
	The contents of VR product are up-to-date enough for my purposes.	
	Overall, the contents in the VR product are very good.	
Telepresence	VR product provides experience as if users forget their immediate environment.	[17, 24]
	VR product provides experience as if users visit virtual world than just seen pictures.	
	VR product provides experience as if users momentarily forget where they are.	
	VR product environment generated in comparison with the "real world" is more real or present.	
	VR product provides experience as if users came back to the "real world" after a journey.	
	Overall, VR product generates environment as if users immerse in it.	
Perceived usefulness	I think VR product is very useful to my life in general.	[13, 28]
	I think VR product provides very useful content to me.	
	I think using VR product improves the quality of the things I do.	
	I think using VR product enhances my effectiveness.	
	I think VR product enables me to accomplish the things I do more quick.	
Perceived enjoyment	I feel using VR product is truly funny.	[28]
	Compared to other devices, I feel using VR product is truly enjoyable.	
	I feel the use of VR product gives me pleasure.	
	I feel the use of VR product makes me feel good.	
	I feel using VR product is enjoyable to me.	

(*continued*)

Table 1. (*continued*)

Variable	Measurements	Reference
Continuous usage intention	If I could, I would like to continue to use VR product.	[6, 23]
	I will frequently use the VR product in the future.	
	I will continue to use VR product for a long time.	
	I am very interested in searching for new information of VR product.	
	I will strongly recommend that others can use VR product.	

4.2   Sampling Design and Data Collection

The questionnaire was designed based on the measurements shown in Table 1. Our sample included Chinese consumers who had at least one experience using virtual reality products. We obtained a total of 444 usable responses through an online survey and SNS in China. Each measurement item used a 5-point Likert scale ranging from 1 (strongly disagree) to 5 (strongly agree). Questionnaire data was analyzed using SPSS 20.0 and Smart PLS 2.0.

4.3   Demographic Characteristics

The demographic characteristics of respondents are shown in Table 2. From the table, we can see that respondents were comprised of 50.7% males and 49.3% females. 66.0% of respondents were aged 20 to 29 years old, while 23.4% were aged 30 to 39 years old. The majority of the consumer interest in using virtual reality products was to watch movies (77.9%) and to play games (55.6%).

Table 2. Demographic Characteristics

Construct		Frequency	Percentage
Gender	Male	225	50.7%
	Female	219	49.3%
Age	<20	17	3.8%
	20–29	293	66.0%
	30–39	104	23.4%
	40–49	27	6.1%
	>50	3	0.7%
Interested Contents (Multiple Choice)	Game	247	55.6%
	Movie	346	77.9%
	Travel	159	35.8%
	Shopping	152	34.2%
	Video chatting	109	24.5%
	Adult contents	65	14.6%
	Others	10	2.3%
Total number = 444			

Table 3. Reliability and Validity

Construct	Items	Factor Loading	Composite Reliability	Cronbach's α	AVE
VR content quality	CON_1	0.746	0.898	0.864	0.596
	CON_2	0.747			
	CON_3	0.774			
	CON_4	0.771			
	CON_5	0.774			
	CON_6	0.819			
Telepresence	TEL_1	0.800	0.907	0.877	0.619
	TEL_2	0.812			
	TEL_3	0.796			
	TEL_4	0.734			
	TEL_5	0.806			
	TEL_6	0.772			
Perceived usefulness	PU_1	0.825	0.902	0.864	0.648
	PU_2	0.763			
	PU_3	0.813			
	PU_4	0.800			
	PU_5	0.823			
Perceived enjoyment	PE_1	0.769	0.902	0.865	0.649
	PE_2	0.807			
	PE_3	0.834			
	PE_4	0.783			
	PE_5	0.834			
Continuous usage intention	CUI_1	0.791	0.900	0.860	0.642
	CUI_2	0.817			
	CUI_3	0.824			
	CUI_4	0.773			
	CUI_5	0.800			

4.4   Reliability and Validity

Table 3 shows the factor loadings, reliability and AVE (Average Variance Extracted) of the measurement model. All factor loadings are above the cutoff value of 0.7. Cronbach's α of each construct exceeds 0.7 and composite reliability of each construct are all higher than 0.7, indicating good reliability of the constructs. Moreover, all AVE values exceed 0.5; that is to say, constructs display convergent validity. Table 4 shows the discriminant validity and all correlation indicators are less than the square root of AVE.

Table 4. Discriminant Validity

	Mean	Std. Dev.	CON	TEL	PU	PE	CUI
CON	3.657	0.701	**0.772**				
TEL	3.746	0.723	0.668	**0.787**			
PU	3.564	0.732	0.735	0.675	**0.805**		
PE	3.853	0.701	0.715	0.766	0.718	**0.806**	
CUI	3.800	0.681	0.691	0.723	0.732	0.786	**0.801**

Note: Diagonal show the square root of the AVE for each construct
CON = VR Content Quality; TEL = Telepresence;
PU = Perceived Usefulness; PE = Perceived Enjoyment;
CUI = Continuous Usage Intention

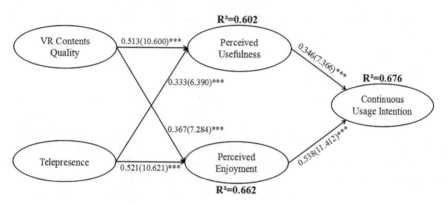

Fig. 2. Hypotheses testing result

4.5 Hypothesis Test

Figure 2 shows the result of the hypotheses tests. Two independent variables have significant positive effects on perceived usefulness and perceived enjoyment. VR content quality has a significant positive influence on perceived usefulness (H1a: 0.513***) and the impact is relatively high when compared with telepresence's influence on perceived usefulness (H2a: 0.333***). VR content quality influence on perceived enjoyment is also supported (H1b: 0.367***), but the impact is relatively small when compared with telepresence's influence on perceived enjoyment (H2b: 0.521***). Continuous usage intention is positively influenced by both perceived usefulness (H3: 0.346***) and perceived enjoyment (H4: 0.538***).

5 Conclusion

5.1 Discussion and Implications

Virtual reality products are new technology products, and thus, there have been few empirical studies related to this domain. This study is among the first empirical studies to examine the continuous usage intention of virtual reality products for daily life. Although few previous studies have investigated virtual reality products for daily life, the results of this study are consistent with prior IT/IS studies in many ways.

First, concerning the influence of consumer intention to continue using virtual reality products, perceived enjoyment has a strong significant effect. This result is consistent with studies by many researchers [12, 18, 26], and suggested in hedonic information system, and intrinsic motivation plays an important role. Virtual reality products, as entertainment products, arouse stronger consumer intention to use it. Therefore, virtual reality product manufactures, content developers, product promoters, and product sellers should try their best to improve products' hedonic value. Second, perceived usefulness positively influences continuous usage intention, which is consistent with findings in [16, 18, 23, 26]. Our research results also support Van der Heijden [25] 's perspective that when customers use hedonic information technology, perceived enjoyment contributes more than perceived usefulness.

Third, content quality positively influences both perceived enjoyment and perceived usefulness. This result is consistent with prior studies that high quality social networking service content helps users obtain more useful and enjoyable experiences [23]. High quality augmented reality content increases both user perception of usefulness and perception of enjoyment [16], and the high quality of media tablet content positively affects both perceived usefulness and enjoyment [28]. Virtual reality products require high quality content. Hence, virtual reality manufactures should attract expert content developer partners and supervise high quality contents and applications. In addition, it is also important to update contents and applications over time. Fourth, telepresence also positively influences both intrinsic and extrinsic motivation, as found in [20, 21]. Telepresence as an important determinant of intrinsic motivation, improves enjoyment and product manufactures should improve their technological level of telepresence implementation (e.g. how to reduce delay and sensations of dizziness, how to reduce distracting sparkle).

5.2 Limitations and Future Research

This study is a rare empirical study of virtual reality products in consumer-oriented daily life. However, this study has some limitations. First, we used content quality and telepresence as independent variables. However, additional variables such as interaction and immersion could be considered. Second, the consumer-oriented virtual reality products have high-price products and low-price products. Different priced virtual reality products feature different virtual reality experiences, different content quality, and different telepresence.

This study also has some implications for future research, and future research can consider the following points. First, investigators should utilize a larger sample size for enhanced representative exploration. Second, as mentioned prior, different priced virtual reality products provide different virtual reality experience, content quality and telepresence. This may influence consumer continuance intention, as high-price product users maybe express more continuance intention than low-price product user, so further study can consider this in their investigation.

References

1. Ahn, J.C., Cho, S.P., Jeong, S.K.: Virtual reality to help relieve travel anxiety. Trans. Internet Inf. Syst. **7**, 1433–1448 (2013)
2. Almaiah, M.A., Jalil, M.A., Man, M.: Extending the TAM to examine the effects of quality features on mobile learning acceptance. J. Comput. Educ. **3**, 453–485 (2016)
3. Bertrand, M., Bouchard, S.: Applying the technology acceptance model to VR with people who are favorable to its use. J. Cyber Therapy Rehabil. **1**, 200–210 (2008)
4. Bowman, D.A., McMahan, R.P.: Virtual reality: how much immersion is enough? Computer **40**, 36–43 (2007)
5. Bryson, S.: Virtual reality in scientific visualization. Commun. ACM **39**, 62–71 (1996)
6. Chang, C.C.: Examining users' intention to continue using social network games: a flow experience perspective. Telematics Inform. **30**, 311–321 (2013)
7. Choi, S., Jung, K., Noh, S.D.: Virtual reality applications in manufacturing industries: past research, present findings, and future directions. Concurrent Eng. **23**, 1–24 (2015)
8. Davis, F.D., Bagozzi, R., Warshaw, P.R.: Extrinsic and intrinsic motivation to use computers in the workplace. J. Appl. Soc. Psychol. **22**, 1111–1132 (1992)
9. Draper, J.V., Kaber, D.B., Usher, H.M.: Telepresence. Hum. Factors **40**, 354–375 (1998)
10. Figueiredo, M., BÖhm, K., Teixeira, J.: Advanced interaction techniques in virtual environments. Comput. Graphics **17**, 655–661 (1993)
11. Guttentag, D.A.: Virtual reality: applications and implications for tourism. Tour. Manag. **31**, 637–651 (2010)
12. Hsiao, C.H., Chang, J.J., Tang, K.Y.: Exploring the influential factors in continuance usage of mobile social apps: satisfaction, habit, and customer value perspectives. Telematics Inform. **33**, 342–355 (2016)
13. Hsu, C.L., Lu, H.P.: Why do people play on-line games: An extended TAM with social influences and flow experience. Inf. Manag. **41**, 853–868 (2004)
14. Huang, H., Rauch, U., Liaw, S.: Investigating learners' attitudes toward virtual reality learning environments: Based on a constructivist approach. Comput. Educ. **55**, 1171–1182 (2010)
15. Jung, T., Dieck, M.C.T., Rauschnabel, P., Ascencao, M., Tuominen, P., Moilanen, T.: Functional, hedonic or social? Exploring antecedents and consequences of virtual reality rollercoaster usage. Augmented Reality and Virtual Reality, pp. 247–258 (2017)
16. Kim, K., Hwang, J., Zo, H., Lee, H.: Understanding users' continuance intention toward smartphone augmented reality applications. Inf. Dev. **32**, 1–14 (2016)
17. Klein, L.R.: Creating virtual product experiences: The role of telepresence. J. Interact. Mark. **17**, 41–55 (2003)
18. Lin, K., Lu, H.: Why people use social networking sites: an empirical study integrating network externalities and motivation theory. Comput. Hum. Behav. **27**, 1152–1161 (2011)
19. Merikivi, J., Tuunainen, V., Nguyen, D.: What makes continued mobile gaming enjoyable? Comput. Hum. Behav. **68**, 411–421 (2017)

20. Park, J.S., Hyun, J., Fairhurst, A., Lee, K.H.: Perception of presence as antecedents to e-tail shopping: an extended technology acceptance model. Res. J. Costume Cult. **20**, 451–461 (2012)
21. Pelet, J., Ettis, S., Cowart, K.: Optimal experience of flow enhanced by telepresence: evidence from social media use. Inf. Manag. **54**, 115–128 (2017)
22. Ryan, R.M., Deci, E.L.: Intrinsic and extrinsic motivations: classic definitions and new directions. Contemp. Educ. Psychol. **25**, 54–67 (2000)
23. Seol, S., Lee, H., Yu, J., Zo, H.: Continuance usage of corporate SNS pages: a communicative ecology perspective. Inf. Manag. **56**, 740–751 (2016)
24. Steuer, J.: Defining virtual reality: dimensions determining telepresence. J. Commun. **42**, 73–93 (1992)
25. van der Heijden, H.: User acceptance of hedonic information systems. MIS Q. **28**, 695–704 (2004)
26. Wamba, S.F., Bhattacharya, M., Trinchera, L., Ngai, E.W.T.: Role of intrinsic and extrinsic factors in user social media acceptance within workspace: Assessing unobserved heterogeneity. Int. J. Inf. Manage. **37**, 1–13 (2017)
27. Whyte, J., Bouchlaghem, N., Thorpe, A., McCaffer, R.: From CAD to virtual reality: modelling approaches, data exchange and interactive 3D building design tools. Autom. Constr. **10**, 43–55 (2000)
28. Yu, J., Lee, H., Ha, I., Zo, H.: User acceptance of media tablets: an empirical examination of perceived value. Telematics Inform. **34**, 206–223 (2017)

A Study on the Method of Removing Code Duplication Using Code Template

Woochang Shin[⊠]

Department of Computer Science, SeoKyeong University, Seoul, Korea
wcshin@skuniv.ac.kr

Abstract. In software development, it is common to use similar code redundantly in many places. However, source code duplication has been reported to adversely affect program quality and maintenance costs. In particular, when writing a program that reflects various conditions, an excessive number of branch ("if-else" or "switch") statements is used; therefore, many of the statements executed for each condition are duplicated. In the present study, we propose a refactoring method for finding duplicate codes used in branch statement and removing them. Based on the proposed method, we also develop and test a prototype tool. The results of the tool test in case studies show that refactoring of the source code written by unskilled developers with the developed tool yields on average 10% reduction in the source code.

Keywords: Code duplication · Branch · Refactoring · Code template

1 Introduction

In software development, it is common to use similar codes redundantly in many places. According to recent estimates, 10–15% of the source code of large software systems is duplicated, and redundant code is not a little portion in medium sized software systems [22,23,36–38]. In general, source code duplication has been reported to adversely affect program quality and maintenance costs [2,12,20,35,40]. Duplication of the source code increases the size and complexity of the program. In addition such duplication is also a sign that software developers are not making good use of program abstraction, such as functions and classes [12,35].

Source code duplication does not simply increase the program size, but also causes subtle errors in the program [29,30]. A "copy-paste" of the code can cause bugs, because it does not properly fix parts that need some modification. For instance, [30] developed the CP-Miner tool, found 28 copy-paste related bugs in the Linux source code, and found 23 related errors in the FreeBSD source code.

Code duplication is reported to be more severe in industrial software systems [1,2,5,11,12,21,31]. It is common to copy and paste source code without a complete understanding of the source code written by other team members. In that case, while the software function will work, the comprehensibility and

R. Lee (Ed.): SNPD 2018, SCI 790, pp. 27–41, 2019.
https://doi.org/10.1007/978-3-319-98367-7_3

maintainability of the software will be compromised [8, 16]. If a bug is found in the duplicate code, all similar codes should be found and corrected [40].

Why do programmers cause code duplication that lead to problems outlined above? [35] provided the following reasons for code duplication:

- Development strategy: Code-clone can be introduced due to the different reuse and programming approach.
- Maintenance benefits: Code-clone can be introduced to obtain several maintenance benefits. (Example: "Risk in developing new code").
- Overcoming underlying limitations: Code-clone can be introduced due to the language/programmer's limitations.
- Cloning by accident: Code-clone can be introduced by accident (Example: "Coincidentally implementing the same logic by different developers").

Among these causes, it is the duplication of codes caused by "Programmer's Limitations" that reveals the problem. Insufficient knowledge of the domain or lack programming skills on part of the programmer lead to the generation of many code-clones [27]. In particular, when writing a program that reflects various conditions, an excessive number of branch ("if-else" or "switch") statements is used, and many of the statements executed for each condition are duplicated. The code clone that occurs in branch statements is underestimated, even though it constitutes a significant part of the source code for a particular program [18]. Excessive use of branch statements increases software complexity with code duplication. The Cyclomatic Complexity software metric proposed by McCabe is proportional to the number of branching of the source code. The larger the Cyclomatic Complexity value, the higher the error probability of the software [14, 32, 43, 44].

In the present paper, we propose a refactoring method to find duplicate codes used in branch statements and removing them. A system model is presented to formalize the problem domain. In the remainder of this paper, we explain the refactoring step to remove redundant code according to the proposed system model. Thereafter, we develop a refactoring tool to remove the code clone with branch and validate the proposed method through case studies.

The remainder of the paper is organized as follows.

In Sect. 2, background and related works are reviewed. In Sect. 3, a system model is presented to formalize the problem domain. Section 4 describes the implementation of the detection and refactoring system with concrete examples. In Sect. 5, we test the effectiveness of the proposed refactoring method with case studies. Section 6 provides the summary of results and draws conclusion of the present study.

2   Background and Related Work

At present, no general definition of code duplication (or code clones) is available. For instance, [5] defines code clones as the segments of code that are "similar" according to some definition of similarity. Furthermore, [37] uses the similarity function to define the same code fragments as the clone group. Likewise, [7] terms a code segment as clone if there is a second or more occurrences of that segment in the source code with or without "minor" modifications. However, as can be seen in the definitions above, terms like "similarity" or "minor change" used to define code clones increase ambiguity. In order to eliminate this ambiguity, [27] classified code-clone types into three categories based on "textual similarity". In addition, [37] added the "functional similarity" category in the code-clone classification of [27] (see below).

- Type 1: A code clone is an exact copy without modifications.
- Type 2: A code clone is a syntactically identical copy; only variable, type, or function identifiers have been changed.
- Type 3: A code clone is a copy with further modifications; statements have been changed, added, or removed.
- Type 4: Two or more code fragments that perform the same computation, but are implemented through dierent syntactic variants.

Numerous studies have been conducted to detect the code clone [34]. Relevant research has also sought to find code clones based on the "text" or "token" of the source code [2,9,18,21,25], and some studies based on the "Abstract Syntax Tree" have also been conducted [5,10]. There has also been research seeking to develop code clone detection algorithms and tools based on software metrics such as [1,19,33].

Compared with the burgeoning research on code clone detection, studies on code clone elimination are scarce. The most common way to eliminate code clone is to use functional abstraction. In such studies, duplicate code is extracted into a new function, and duplication of code is eliminated by calling this function from the existing code [13,26,42]. Design patterns have been applied to remove redundant codes at the design level [3,4].

Several studies have attempted to remove branch statements regardless of code duplication [6,28]. Typically, in these studies, branch statements were refactored using polymorphism of object-oriented programming [15,41]. Null object pattern has been proposed to remove conditional statements that check the existence of the object [17]. However, in the substantial research on code clones, few studies have managed to effectively eliminate code clones in branch statements. In this context, the present study seeks to make the following contributions to the field:

1. A proposal for a refactoring method to search and remove code clones in branch statements.
2. Development of a refactoring tool to eliminate code clones in branch statements.
3. Validation of the proposed refactoring method through case studies.

3 System Model for Refactoring

The type of code clones that is subject to refactoring in the present study is Type 2. Statement $S1$ is a clone of another statement $S2$, if $S1$ is syntactically identical to $S2$ except for variations in identifiers, literals, and expressions. In this paper, the following terms and function definitions are used:

Definition 1 (Statement). A statement S is a tuple constructed as specified below.

$$S = \ <stype, ExprSeq, CTX>$$

- *stype*: statement type
  - $stype \in \{ST\_ASSIGN, ST\_IF, ST\_FOR, ST\_VAR\_DECL, ...\}$
- *ExprSeq*: a sequence of expressions $<expr_1, ..., expr_n>$, $expr_1, ..., expr_n \in EXPRS$
  - *EXPRS*: a set of all expressions
- CTX: execution context containing information about variables and functions.

Definition 2 (Expression). An expression EX is a tuple constructed as specified below.

$$EX = \ <etype, ExprSeq, CTX>$$

- *etype*: expression type
  - $etype \in \{ET\_VAR, ET\_FUNC\_CALL, ET\_PLUS, ET\_MINUS, ...\}$
- *ExprSeq*: a sequence of expressions $<expr_1, ..., expr_n>$, $expr_1, ..., expr_n \in EXPRS$
- CTX: execution context containing information about variables and functions.

Definition 3 (Code Fragment). A code fragment CF is a sequence of statements

$$CF = \ <s_1, s_2, ..., s_n>,   s_1, s_2, ..., s_n \in STMTS$$

- *STMTS*: a set of all statements

Figure 1 is a part of the source codes written by a computer science student in a semester project.

According to Definitions 1, 2, and 3, each conditional expression and execution statements is represented as follows.

Conditional Expressions: "mapnum == 1", "mapnum == 2", "mapnum == 3", "mapnum == 4"

- $cond\_expr1 = \ < ET\_EQUAL, << ET\_VAR, "mapnum", ctx >,$
  $< ET\_LITERAL\_INT, 1, ctx >>, ctx >$
- $cond\_expr2 = \ < ET\_EQUAL, << ET\_VAR, "mapnum", ctx >,$
  $< ET\_LITERAL\_INT, 2, ctx >>, ctx >$
- $cond\_expr3 = \ < ET\_EQUAL, << ET\_VAR, "mapnum", ctx >,$
  $< ET\_LITERAL\_INT, 3, ctx >>, ctx >$
- $cond\_expr4 = \ < ET\_EQUAL, << ET\_VAR, "mapnum", ctx >,$
  $< ET\_LITERAL\_INT, 4, ctx >>, ctx >$

```
if ( mapnum == 1 ) {
    ranktime[0] = 0.0; change_map(1);
} else if ( mapnum == 2 ) {
    ranktime[1] = 0.0; change_map(2);
} else if ( mapnum == 3 ) {
    ranktime[2] = 0.0; change_map(3);
} else if ( mapnum == 4 ) {
    ranktime[3] = 0.0; change_map(4);
} else {
    ranktime[4] = 0.0; change_map(5);
}
```

Fig. 1. Sample code

Execution Statement 1: "ranktime[0] = 0.0;"

- $exec\_stmt1\_1 = <ST\_ASSIGN, <lexpr, rexpr>, ctx>$
  - $lexpr = <ET\_ARR\_ELM, <<ET\_ARR\_NAME,"ranktime", ctx>,$
    $<ET\_ARR\_INDEX, <<ET\_LITERAL\_INT, 0, ctx>>$
    $, ctx>>, ctx>$
  - $rexpr = <ET\_LITERAL\_DOUBLE, 0.0, ctx>$

Execution Statement 2: "change_map(1);"

- $exec\_stmt1\_2 = <ST\_FUNC\_CALL, <func\_name, func\_params>, ctx>$
  - $func\_name = <ET\_FUNC\_NAME,"change\_map", ctx>$
  - $func\_params = <ET\_FUNC\_PARAMS, <<ET\_LITERAL\_INT,$
    $1, ctx>>, ctx>$

In Fig. 1, it can be seen that the conditional expressions and execution statements are duplicated. In order to remove code duplication, a template is created by extracting common parts from duplicate expressions or statements. The definition of the template is as follows:

Definition 4 (Expression Template). Expression template ET is a tuple consisting of an expression and a reference sequence to the sub-expressions to be replaced by arguments.

$$ET = <expr, REFS>$$

- $expr$: an expression containing parameter-markers. $expr \in EXPRS$
- $REFS$: a reference sequence for the sub-expressions to be replaced by arguments in $expr$. $= <ref_1, ref_2, \ldots, ref_n>$

Definition 5 (Code Template). Code template CT is a tuple consisting of a code fragment and a reference sequence to the sub-expressions to be replaced by an argument.

$$CT = <cf, REFS>$$

- cf: a code fragment containing parameter-markers.
- $REFS$: a reference sequence for the sub-expressions to be replaced by arguments in $cf. = <ref_1, ref_2, \ldots, ref_n>$

In Fig. 1, the four conditional expressions - $cond\_expr1$, $cond\_expr2$, $cond\_expr3$, $cond\_expr4$ - are identical except for the integer values to be compared. In these expressions, we replace the different parts with parameter markers and extract common parts to make an expression template as follows:

$$sample\_cond\_template = <<ET\_EQUAL, <<ET\_VAR," mapnum", ctx>,$$
$$pmarker>, ctx>, refs>$$

- $pmarker$: a parameter marker. $= <ET\_MARKER, <>, ctx>$
- $refs = <\&pmarker>$

The parameter marker indicates the location of the expression to be replaced with a parameter in the template. When source code is generated by applying the template, parameters as many as the number of parameter markers are required. In Fig. 1, comparing the execution statements, $cf1$="ranktime[0] = 0.0; change_map(1);", $cf2$="ranktime[1] = 0.0; change_map(2);", $cf3$="ranktime[2] = 0.0; change_map(3);", $cf4$="ranktime[3] = 0.0; change_map(4);", and $cf5$="ranktime[4] = 0.0; change_map(5);", for each condition, all but the two integer values are identical. A code template $sample\_exec\_template$ is created by extracting the common parts of $cf1$, $cf2$, $cf3$, $cf4$, and $cf5$.

$$sample\_exec\_template = <<tstmt1, tstmt2>, refs>$$

- $tstmt1 = <ST\_ASSIGN, <lexpr, rexpr>, ctx>$
  - $lexpr = <ET\_ARR\_ELM, <<ET\_ARR\_NAME," ranktime", ctx>,$
    $<ET\_ARR\_INDEX, <pmarker1>, ctx>>, ctx>>, ctx>$
  - $rexpr = <ET\_LITERAL\_DOUBLE, 0.0, ctx>$
- $tstmt2 = <ST\_FUNC\_CALL, <func\_name, func\_params>, ctx>$
  - $func\_name = <ET\_FUNC\_NAME," change\_map", ctx>$
  - $func\_params = <ET\_FUNC\_PARAMS, <pmarker2>, ctx>$
- $pmarker1, pmarker2$: parameter markers. $= <ET\_MARKER, <>, ctx>$
- $refs = <\&pmarker1, \&pmarker2>$

A template represents a common part of duplicate expressions or redundant code fragments; however, it cannot generate a source code by itself. That is, the parameters are needed to generate source code from the template. The relationship between redundant expressions, a template, and parameters are shown in Fig. 2.

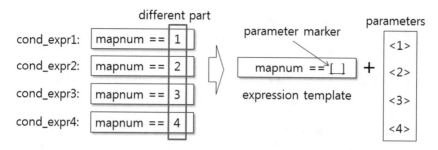

Fig. 2. Creating an expression template from duplicate expressions

As shown in Fig. 2, the identical parts in duplicated codes are made into templates, and the different parts are stored as parameters. The definition of the template generation function TGF that performs this task is as follows:

Definition 6 (Template Generation Function). The function TGF that creates a template from expressions or code fragments is provided below.

[From expressions]

$$TGF(exprs) : EXPR\_SEQ \rightarrow EXPR\_TEMPLATE \times PARAMS\_SEQ$$

- $TGF(exprs) = \ <expr\_template, params\_seq>$ where $exprs = \ <expr_1, expr_2, \ldots, expr_n>$
- $expr\_template$: Expression template that represents the common part of all expressions in $exprs. = \ <texpr, REFS>$
- $params\_seq$: a sequence of parameters. $= \ <params_1, params_2, \ldots, params_n>$
  - $params_i$: The sequence of expressions corresponding to the $pmarker$ position of $expr\_template$ in $expr_i. = \ <p_i\_expr_1, p_i\_expr_2, \ldots, p_i\_expr_m>$

[From code fragments]

$$TGF(cfs) : CODE\_FRAGMENTS \rightarrow CODE\_TEMPLATE \times PARAMS\_SEQ$$

- $TGF(cfs) = \ <code\_template, params\_seq>$ where $cfs = \ <cf_1, cf_2, \ldots, cf_n>$
- $code\_template$: A code template that represents the common parts of all code fragments in $cfs. = \ <tcode, REFS>$
- $params\_seq$: a sequence of parameters. $= \ <params_1, params_2, \ldots, params_n>$
  - $params_i$: The sequence of expressions corresponding to the $pmarker$ position of $code\_template$ in $cf_i. = \ <p_i\_expr_1, p_i\_expr_2, \ldots, p_i\_expr_m>$.

Apply the template generation function TGF to the condition expressions $cond\_expr1$, $cond\_expr2$, $cond\_expr3$, and $cond\_expr4$ in the sample code. The result is shown below.

$$TGF(< cond\_expr1, cond\_expr2, cond\_expr3, cond\_expr4 >) = < sample\_cond\_$$
$$template, << param_1 >, < param_2 >, < param_3 >, < param_4 >>>$$

- $sample\_cond\_template = << ET\_EQUAL, << ET\_VAR," mapnum", ctx >,$ $pmarker >, ctx >, < \&pmaker >>$
- $param_i = < ET\_LITERAL\_INT, i, ctx >$, where $1< = i <= 4$

To generate a template using TGF, the following prerequisites should be satisfied.

[Prerequisites]

1. When creating a template, variables used in different parts of the expressions should not be used as l-values. For example, when comparing two statements "a = 1;" and "b = 1;", you cannot create a template, because the different parts (a, b) are on the left of the assignment.
2. In multiple code fragments, all variables used in the expressions that differ from each other should not be changed from the first statement of the template to the statement of the corresponding expression.
   - The different parts of the two code fragments "a = 2; b = a;" and "a = 2; b = 3;" are 'a' and '3' in the second statement. The variable 'a' does not satisfy this condition, because its value changes to 2 in the first statement. That is, you cannot create a template.

The function TAF, which generates the code from a template, is defined below.

Definition 7 (Template Application Function). The TAF function that generates source code by applying the parameters to the template is defined as follows:

[From expression template]

$$TAF(expr\_template, params) : EXPR\_TEMPLATE \times PARAMS \rightarrow EXPRESSION$$

- $TAF(expr\_template, params) = < etype, < expr_1, \ldots, expr_m >, ctx >$
- $expr\_template$: an expression template. $= << tetype, < texpr_1, \ldots, texpr_m >, ctx >, < ref_1, \ldots, ref_n >>$
- $params$: a sequence of parameter expressions. $= < pexpr_1, pexpr_2, \ldots, pexpr_n >$
- $etype = tetype$
- $expr_i = texpr_i[pexpr_1/ref_1, pexpr_2/ref_2, \ldots, pexpr_n/ref_n]$, where $1< = i <= m$

- $expr[pexpr/ref]$: If any of the sub-expressions that constitute $expr$ contain the expression pointed to by ref, replace it with the specified $pexpr$. If $expr$ does not have the expression pointed to by ref, this operation is ignored.

[From code template]

$$TAF(code\_template, params) : CODE\_TEMPLATE \times PARAMS \rightarrow CODE\_FRAG$$

- $TAF(code\_template, params) = \; < stmt_1, \ldots, stmt_m >$
- $code\_template$: a code template. $= \; < \; < \; tstmt_1, tstmt_2, \ldots, tstmt_m >, < ref_1, \ldots, ref_n >>$
- $params$: a sequence of parameter expressions. $= \; < pexpr_1, pexpr_2, \ldots, pexpr_n >$
- $stmt_i = tstmt_i[pexpr_1/ref_1, pexpr_2/ref_2, \ldots, pexpr_n/ref_n]$, where $1 < = i <= m$
  - $stmt[pexpr/ref]$: If any of the expressions that constitute a statement $stmt$ are found by ref, replace it with the specified $pexpr$. If the expression pointed to by ref is not in $stmt$, this operation is ignored.

If you have created the template $sample\_cond\_template$ for the conditional expressions $cond\_expr1$, $cond\_expr2$, $cond\_expr3$, and $cond\_expr4$ in the sample code, you can recreate $cond\_expr1$ by applying this template to the TAF function.

$$TAF(sample\_cond\_template, <<ET\_LITERAL\_INT, 1, ctx >>) = cond\_expr1$$

4   Refactoring Method and Implementation

In the present study, we developed a prototype tool that can find code duplication used in branch statements and refactor the code. The development environment of the tools is as follows.

- Java Development Kit v1.8
- Eclipse Neon 2
- JavaSymbolSolver v0.5.1 [39]

The procedure for refactoring to remove a code-clone with branch is outlined below.

Step 1. Detect code-clones with branch.

(a) Convert the source code into Abstract Syntax Tree (AST) format.
(b) Search for branch ("if" or "switch") statement in AST.
(c) In the branch statement, analyze the execution code fragments for each condition and check whether the code template can be generated. If you can create a code template, proceed to the next step.

Step 2. Make templates (conditional expression template and execution code template).

(a) Classify conditional expressions and execution code fragments in the branch statement.
(b) Create an expression template from conditional expressions and store the different parts as parameter *cond_params* (optional).
(c) Create a code template from execution code fragments and store the different parts as parameter *cf_params* (mandatory).

Step 3. Generate refactored code.

(a) The refactored code consists of the conditional part and the execution part.
(b) In the conditional part, find the index of the first conditional expression that is true when you evaluate the conditional expressions in order.
   – If an expression template is created for the conditional expressions, you can evaluate each conditional expression by applying the *cond_params* stored in Step 2 to this template.

```
int index=0;
Object[][] cond_params = { ... };
for( ; index < cond_params.length ; ++index)
    if ( TAF(expression_template, cond_params[index]) )
        break;
```

   – If no expression template is created for the conditional expressions, each conditional expression is converted to a separate JAVA lambda function and then is evaluated.
(c) In the execution part, using the index found in the conditional part, find the parameters (*cf_params[index]*) and generate source code by applying these parameters and the code template to the *TAF* function.

```
TAF(code_template, cf_params[index]);
```

   – If the end of the branch statement is "else" (or "default" if it is a switch statement), no additional condition check is required to execute the code of the execution part. If not, it is necessary to check whether there is a conditional expression that is true.

```
if ( index == cond_params.length )
    TAF(code_template, cf_params[index]);
```

When the code shown in Fig. 1 is refactored by the proposed method, in the following results are obtained (see Fig. 3).

In Fig. 3, lines 1 to 4 are conditional, while lines 5 to 7 are execution parts. *cond_params* stores different parts of the conditional expressions. The expression "mapnum == cond_params [index]" on line 4 is the result of applying template *expression_template*. Line 5 defines the parameters of the code template, while lines 6 through 7 are code generated by applying the *code_template* and parameters to the *TAF* function. The source code generated by applying the refactoring step described above can be further optimized in some cases. If the conditional expressions are constant value comparisons for one variable, and the values to be compared form an arithmetic sequence, then the conditional code can be simplified as follows:

$$index = (compared\_variable - first\_term\_of\_sequence)/common\_difference$$

```
01: int index=0;
02: int cond_params[] =  1, 2, 3, 4 ;
03: for( ; index < cond_params.length ; ++index)
04:      if ( mapnum == cond_params[index] ) break;
05: int[][] cf_params = { {0, 1}, {1, 2}, {2, 3}, {3, 4}, {4, 5} };
06: ranktime[ cf_params[index][0] ] = 0.0;
07: change_map( cf_params[index][1] );
```

Fig. 3. Refactored sample code

If you can directly calculate the *cf_params* values using the index of the conditional part, you can apply the calculation formula directly to the *execution_template* without *cf_params*. The refactored sample code with the optimization method is shown in Fig. 4.

```
int index = (0 <= (mapnum-1) && (mapnum-1) <=3 ) ? (mapnum-1) : 4;
ranktime[ index ] = 0.0;
change_map( index+1 );
```

Fig. 4. Refactored sample code (with optimization)

5   Case Studies

The results of the measurement of the Cyclomatic Complexity proposed by McCabe (the software metric that measures the complexity of the program) for the original sample code (Fig. 2) and the refactored source codes (Figs. 3 and 4) are shown in Table 1.

As a result of refactoring, as compared with the original code, the refactored code for LOC, LLOC, and Cyclomatic Complexity decreased by 56.3%, 40%, and 40%, respectively. In addition, as compared to the original sample code the refactored code with optimization decreased by 81.3% of LOC, 75% of LLOC, and 80% of Cyclomatic Complexity. Therefore, we can reduce the size

Table 1. Quantitative comparison of the original and refactored source codes

	Original sample code	Refactored sample code	Refactored code with optimization
Line of Code (LOC)	16	7	3
Logical Line of Code (LLOC)	20	12	5
Cyclomatic Complexity	5	3	1

and complexity of the source code by using the refactoring method proposed in the present study for branch statements with redundant code.

For the case study, we analyzed the source code submitted to the semester project (game development) of the computer science students' programming class using the tool developed in the present study. The results are shown in Table 2.

Table 2. Effects of refactoring the source code written by unskilled developers

	File number (A)	Total LOC (A)	Branch number	Branch LOC	TGF-applicable branch number	TFG-applicable branch LOC (B)	B/A	Line of refactored code (C)	(B-C)/A
Team A	3	723	48	410	5	193	26.7%	92	14.0%
Team B	11	1514	273	813	17	122	8.1%	52	4.6%
Team C	5	881	49	562	25	291	33.0%	125	18.8%
Team D	9	2108	94	1493	18	282	13.4%	139	6.8%
Team E	3	791	24	479	7	175	22.1%	85	11.4%
Team F	7	881	17	228	3	40	4.5%	13	3.1%

Depending on the students level of programming skills, there was a large difference in the ratio of using branch statements and the rate of code duplication. The percentage of code clone with branch in the project's source code ranged from 4.5% to 33.0%. As a result of eliminating code redundancy in branch statements using the tools developed in the present study, the ratio of the reduced source code ranged between 3.1% and 18.8%. Therefore, on average, ca. 10% of the total source code is reduced.

6   Conclusion

In the present paper, we introduced a refactoring method that detects and removes code-clones with branch. The problem domain is expressed using the formal system model, and each step of the refactoring method is explained. We developed a prototype tool to remove code clones with branch according to the proposed method, and validated the proposed method through case studies. The results of refactoring the source code written by unskilled developers with the developed tool yielded on average 10% reduction in the source code. In future research, developing models and tools for search and elimination of code clone type-3 in branch statements would be needed.

Acknowledgements. This Research was supported by Seokyeong University in 2017.

References

1. Antoniol, G., Villano, U., Merlo, E., Penta, M.D.: Analyzing cloning evolution in the Linux kernel. Inf. Softw. Technol. **44**(13), 755–765 (2002)
2. Baker, B.S.: On finding duplication and near-duplication in large software systems. In: Proceedings of the Second Working Conference on Reverse Engineering (WCRE 1995). IEEE Computer Society, Washington, DC (1995)
3. Balazinska, M., Merlo, E., Dagenais, M., Lague, B., Kontogiannis, K.: Partial redesign of java software systems based on clone analysis. In: Working Conference on Reverse Engineering, pp. 326–336. IEEE Computer Society Press (1999)
4. Balazinska, M., Merlo, E., Dagenais, M., Lague, B., Kontogiannis, K.: Advanced clone-analysis to support object-oriented system refactoring. In: Working Conference on Reverse Engineering, pp. 98–107. IEEE Computer Society Press (2000)
5. Baxter, I.D., Yahin, A., Moura, L., Sant'Anna, M., Bier, L.: Clone detection using abstract syntax trees. In: Proceedings of the International Conference on Software Maintenance (ICSM 1998). IEEE Computer Society, Washington, DC (1998)
6. Bodk, R., Gupta, R., Soffa, M.L.: Interprocedural conditional branch elimination. In: Proceedings of the ACM SIGPLAN 1997 conference on Programming Language Design and Implementation, Las Vegas, Nevada, USA, pp. 146–158 (1997)
7. Burd, E., Bailey, J.: Evaluating clone detection tools for use during preventative maintenance. In: Proceedings of the 2nd IEEE International Workshop on Source Code Analysis and Manipulation (SCAM 2002), Montreal, Canada, pp. 36–43 (2002)
8. Cordy, J.R.: Comprehending reality: practical challenges to software maintenance automation. In: Proceedings of the 11th IEEE International Workshop on Program Comprehension (IWPC 2003), Portland, Oregon, USA, pp. 196–206 (2003)
9. Cordy, J.R., Dean, T.R., Synytskyy, N.: Practical language-independent detection of near-miss clones. In: Proceedings of the 14th IBM Centre for Advanced Studies Conference (CASCON 2004), Toronto, Ontario, Canada, pp. 1–12 (2004)
10. Corazza, A., Martino, S.D., Maggio, V., Scanniello, G.: A tree kernel based approach for clone detection. In: Proceedings of the 26th IEEE International Conference on Software Maintenance (ICSM 2010), Timisoara, Romania, pp. 1–5 (2010)
11. Dagenais, M., Merlo, E., Lague, B., Proulx, D.: Clones occurrence in large object oriented software packages. In: Proceedings of the 8th IBM Centre for Advanced Studies Conference (CASCON 1998), Toronto, Ontario, Canada, pp. 192–200 (1998)
12. Ducasse, S., Rieger, M., Demeyer, S.: A language independent approach for detecting duplicated code. In: Proceedings of the IEEE International Conference on Software Maintenance (ICSM 1999). IEEE Computer Society, Washington, DC (1999)
13. Fanta, R., Rajlich, V.: Removing clones from the code. J. Softw. Maint. Evol. **11**(4), 223–243 (1999)
14. Fenton, N.E., Neil, M.: A critique of software defect prediction models. IEEE Trans. Softw. Eng. **25**(3), 675–689 (1999)
15. Fowler, M.: Refactoring: Improving the Design of Existing Code. Addison-Wesley, Boston (1999). ISBN: 0-201-48567-2
16. Giesecke, S.: Generic modelling of code clones. In: Proceedings of Duplication, Redundancy, and Similarity in Software, Dagstuhl, Germany (2006). ISSN 16824405
17. Grand, M.: Patterns in Java: A Catalog of Reusable Design Patterns Illustrated with UML. Wiley, Indianapolis (1998)

18. Higo, Y., Kamiya, T., Kusumoto, S., Inoue, K.: Method and implementation for investigating code clones in a software system. Inf. Softw. Technol. **49**, 985–998 (2007)
19. Hummel, B., Juergens, E., Heinemann, L., Conradt, M.: Index-based code clone detection: incremental, distributed, scalable. In: Proceedings of the 2010 IEEE International Conference on Software Maintenance (ICSM 2010), pp. 1–9. IEEE Computer Society, Washington (2010)
20. Juergens, E., Deissenboeck, F., Hummel, B., Wagner, S.: Do code clones matter? In: Proceedings of the 31st International Conference on Software Engineering (ICSE 2009), pp. 485–495. IEEE Computer Society, Washington, DC (2009)
21. Kamiya, T., Kusumoto, S., Inoue, K.: CCFinder: a multilinguistic token-based code clone detection system for large scale source code. IEEE Trans. Softw. Eng. **28**(7), 654–670 (2002)
22. Kapser, C.J., Godfrey, M.W.: Aiding comprehension of cloning through categorization. In: Proceedings of 2004 International Workshop on Principles of Software Evolution (IWPSE 2004), pp. 85–94 (2004)
23. Kapser, C.J., Godfrey, M.W.: Supporting the analysis of clones in software systems. J. Softw. Maint. Evol. Res. Pract. **18**(2), 61–82 (2006)
24. Kapser, C.J., Godfrey, M.W.: "Cloning considered harmful" considered harmful: patterns of cloning in software. Empir. Softw. Eng. **13**(6), 645–692 (2008)
25. Kawaguchi, S., Yamashina, T., Uwano, H., Fushida, K., Kamei, Y., Nagura, M., Iida, H.: SHINOBI: a tool for automatic code clone detection in the IDE. In: Conference: 16th Working Conference on Reverse Engineering, WCRE 2009, Lille, France, pp. 313–314 (2009)
26. Komondoor, R., Horwitz, S.: Eliminating duplication in source code via procedure extraction. Technical report 1461, UW-Madison Department of Computer Sciences (2002)
27. Koschke, R.: Survey of research on software clones. Duplication, Redundancy, and Similarity in Software (2006)
28. Kreahling, W., Whalley, D., Bailey, M., Yuan, X., Uh, G.R., van Engelen, R.: Branch elimination via multi-variable condition merging. In: Kosch, H., Böszörményi, L., Hellwagner, H. (eds.) Euro-Par 2003 Parallel Processing, Euro-Par 2003. Lecture Notes in Computer Science, vol. 2790. Springer, Heidelberg (2003)
29. Li, Z., Lu, S., Myagmar, S., Zhou, Y.: CP-Miner: a tool for finding copy-paste and related bugs in operating system code. In: Proceedings of the 6th Symposium on Operating System Design and Implementation (OSDI 2004), San Francisco, CA, USA, pp. 289–302 (2004)
30. Li, Z., Lu, S., Myagmar, S., Zhou, Y.: CP-Miner: finding copy-paste and related bugs in large-scale software code. IEEE Trans. Softw. Eng. **32**(3), 176–192 (2006)
31. Mayrand, J., Leblanc, C., Merlo, E.: Experiment on the automatic detection of function clones in a software system using metrics. In: Proceedings of the 12th International Conference on Software Maintenance (ICSM 1996), Monterey, CA, USA, pp. 244–253 (1996)
32. McCabe, T.J.: A complexity measure. IEEE Trans. Softw. Eng. **2**(4), 308–320 (1976)
33. Perumal, A., Kanmani, S., Kodhai, E.: Extracting the similarity in detected software clones using metrics. In: Proceedings of International Conference on Computer and Communication Technology, Allahabad, Uttar Pradesh, India, pp. 575–579 (2010)

34. Rattan, D., Bhatia, R.K., Singh, M.: Software clone detection: a systematic review. Inf. Softw. Technol. **55**, 1165–1199 (2013)
35. Roy, C.K., Cordy, J.R.: A survey on software clone detection research. School of Computing TR 2007-541, Queens University 115 (2007)
36. Roy, C.K., Cordy, J.R.: An empirical study of function clones in open source software systems. In: Proceedings of the 15th Working Conference on Reverse Engineering, WCRE 2008, pp. 81–90 (2008)
37. Roy, C.K., Cordy, J.R., Koschke, R.: Comparison and evaluation of code clone detection techniques and tools: a qualitative approach. Sci. Comput. Program. **74**(7), 470–495 (2009)
38. Schwarz, N., Lungu, M., Robbes, R.: On how often code is cloned across repositories. In: 34th International Conference on Software Engineering New Ideas and Emerging Results Track (ICSE), pp. 1289–1292. IEEE (2012)
39. Smith, N., van Bruggen, D., Tomassetti, F.: JavaParser: Visited (2017). https://leanpub.com/javaparservisited
40. Toomim, M. , Begel, A., Graham, S.L.: Managing duplicated code with linked editing. In: Proceedings of the 2004 IEEE Symposium on Visual Languages - Human Centric Computing (VLHCC 2004), pp. 173–180. IEEE Computer Society, Washington, DC (2004)
41. Tsantalis, N., Chatzigeorgiou, A.: Identification of refactoring opportunities introducing polymorphism. J. Syst. Softw. **83**(3), 391–404 (2010)
42. Tsantalis, N., Chatzigeorgiou, A.: Identification of extract method refactoring opportunities for the decomposition of methods. J. Syst. Softw. **84**, 1757–1782 (2011)
43. Yu, L., Mishra, A.: Experience in predicting fault-prone software modules using complexity metrics. J. Qual. Technol. Quant. Manag. **9**, 421–434 (2012)
44. Zuse, H.: Software Complexity, Measures and Methods. Walter de Gruyter, Berlin, New York (1991)

ICBM-Based Smart Farm Environment Management System

Meonghun Lee[1], Haengkon Kim[2], and Hyun Yoe[3(✉)]

[1] Department of Agricultural Engineering,
National Institute of Agricultural Sciences,
Wanju, Jeollabuk-do 55365, Republic of Korea
leemh5544@gmail.com
[2] School of Information Technology, Catholic University of Daegu,
Republic of Korea, Gyeongsan, Gyeongbuk-do 38430, Republic of Korea
hangkon@cu.ac.kr
[3] Department of Information and Communication Engineering,
Sunchon National University, Suncheon,
Jeollanam-do 57922, Republic of Korea
yhyun@sunchon.ac.kr

Abstract. The combination of Internet of Things (IoT), cloud computing, big data, and mobile technologies is a new technology paradigm referred to as ICBM technology. In this study, we designed and implemented a smart farm environment management system based on the ICBM paradigm that can collect and monitor information on crop growth. Our proposed wireless system not only collects environmental data from inside a greenhouse and controls the greenhouse facilities, but also enhances energy efficiency through effective management of IoT sensor network topology. Our system provides convenience by allowing remote monitoring and controlling of the smart farm environment while establishing a database that enables big data analysis in the cloud to optimize the environment for crop growth. The safety of all functions related to information collection, information delivery, and smart farm control by the user have been confirmed through application of this technology on fields. Furthermore, our proposed system also grants flexibility of time and location when it comes to monitoring and controlling farms or greenhouses.

Keywords: Internet of things · Cloud computing · Big data · Mobile Agriculture · Smart farm

1 Introduction

The global agricultural industry is facing considerable challenges. First, agricultural competition is becoming increasingly fierce following the opening of global agricultural markets, leading to increasing concerns of a possible collapse of the agricultural management base. Second, in contrast to the global competition, the lack of local agricultural competition caused by a drop in rural population and aging society, along with the issue of outdated technology and equipment due to a lack of financial support, are also negative factors affecting the global agricultural industry.

© Springer Nature Switzerland AG 2019
R. Lee (Ed.): SNPD 2018, SCI 790, pp. 42–56, 2019.
https://doi.org/10.1007/978-3-319-98367-7_4

Furthermore, contrary to consumers' expectations for safe food products, the illegal retail sale of low-quality agricultural products is threatening people's health; this highlights the need for advancement of agricultural practices and equipment to ensure high-quality agricultural produce. In this light, the integration of information and communications technology (ICT) with agriculture will not only lead to clean agricultural products, but also contribute greatly to an increase in rural income.

Thus, ICT will enable the systematic management and analysis of both agricultural cultivation and farming data in order to enhance quality and productivity, which will raise the marketable quality of agricultural products by establishing a database for production and retail information, including product origin, seeding period, use of agricultural pesticides, and shipping date. Moreover, establishing an exchange system for sharing information on optimal production conditions for agricultural production (e.g., temperature, humidity, and illumination), details (e.g., seeding period and shipping date), and agricultural market conditions can contribute considerably to the increase in rural income.

If such ICT is used in a farm, it is referred to as a "smart farm." Thus, a smart farm can be established through a platform that unifies Internet of Things (IoT), cloud, big data, and mobile (ICBM) technologies [1–3]; in particular, the integration of these four technologies in diverse industrial areas to develop smart systems has given rise to a new platform called the ICBM platform. Considering this, an investigation of these technologies can provide an outlook on smart agriculture.

2   Background

2.1   IoT Technology

IoT is a technology that integrates the Internet with hardware and information [4]; in particular, it utilizes sensors of both network communication and information technology (IT), which are applied to devices and units ranging from household electronics to a building or an entire city to provide convenience services to users. IoT is similar to machine-to-machine (M2 M) communication; however, they are two distinct concepts [5]. The primary difference is that M2 M communication is defined as communication or information exchange between sensors and devices that does not require human intervention, whereas IoT might allow human input aside from communication and information exchange between sensors and devices. Thus, though, in the past, conventionally only desktop computers were used to exchange information, in modern times, all devices, including smartphones, smart watches, and even information applications, allow communication between objects and amongst themselves.

IoT is fast emerging as the next-generation ubiquitous technology promoting growth in all industrial fields, including agriculture and horticulture [6]. IoT acts as a substitute for tasks and fields where human intervention is not always possible or is difficult, including water supply automation and weather forecasting; furthermore, it is predicted that IoT will provide the general public with easier access to agriculture and horticulture information. In particular, IoT does not only provides environmental data, but also delivers information on crops; it identifies the state of the crops and provides helpful information, such as the amounts of water and illumination required. Moreover, it can also group all the signals sent from different sensors into one application in order to direct sprinklers and automatically control the amount of water and illumination that is supplied to the crops. Therefore, experts have praised IoT as a technology that has reduced costs and increased convenience in the fields of agriculture and horticulture. As an example of its usefulness in agriculture, by supplying crops with the appropriate amount of water at the appropriate times, IoT is expected to decrease the national water usage by 30%.

2.2   Cloud Technology

Cloud computing is a computing paradigm that enables remote connection and usage of shared computing resources through a network [7, 8]. Using cloud computing, users can access programs and data installed and saved on different PCs or devices; in contrast, conventionally, only a limited number of users could access the programs and data saved on a large-scale computer. Thus, cloud computing is a user-oriented computing technology that enables users to perform tasks remotely through a network without any limitations on time and space. In addition, using cloud computing, the requirements of performing a specific task is no longer limited to available software, but can be extended to various other IT resources (e.g., storage, server, network, development platform) [9].

The advantage of cloud technology is that it enables multiple users to simultaneously use a specific database or program from a single server; this is considerably efficient in terms of management because only a single license or one-time installation of a program is required. Moreover, from the perspective of an organization, this shared access can contribute to work efficiency because users can share data by saving them onto the server. In contrast, the current storage technology suffers from its limited capacity to save data and enable communication between devices. Furthermore, because all data are saved onto the server, it leads to more secure data, which serves as an advantage; therefore, in this case, the risk of data leaks is lower than in the case of conventional data management.

The advancement of cloud technology enables analysis of patterns of device usage by saving them onto the remote server; such analysis can be used to investigate device malfunctions, evaluate energy consumption for cooling a greenhouse, or optimize environmental control within a smart farm. In particular, IoT can be used to save data from numerous device sensors in a cloud to conduct big data analysis to provide quality service to the users.

2.3   Big Data

With the advancement of social network services (SNSs) and network technologies, the advent of the IoT era has produced unprecedented data in terms of both quality and quantity [10, 11]. In particular, the increase in the number of SNS users has led to a considerable part of server storage being used up for conventional unstructured data; considering this, the development of technologies that could extract valuable information by analyzing such data is important. Big data addresses this requirement; it does not simply refer to the possession of incomparably massive data, instead, it refers to technologies that can process and analyze such unstructured data. Therefore, big data is promising not only for corporations within specific industries that generate large amounts of data, but also a majority of countries that use modern devices, such as smartphones. The inherent value of big data is to create new value for users (e.g., the information obtained through big data can be used to offer new services earlier because big data technology helps corporations or organizations secure transparency while meeting consumer needs in a timely manner); automated data analysis is necessary to identify such opportunities.

Another crucial function of big data is to disperse risks by easily identifying unusual details and providing them to the users, e.g., identifying issues in crop growth analysis or management of species diversity; however, to realize these reduced risks, organizational change is necessary. In addition to these functions, big data technology can be utilized in a variety of other ways, including optimization of the growth environment, optimization of business management, and marketing.

2.4   Mobile Technology

The advancement of mobile technology has enabled users to use mobile devices for functions that were conventionally performed by computers. Aside from these functions, mobile technology can be used to exchange data independent of time and place, and can make use of real-time GPS information, SNS incoming messages, and even health information [12]. Smart devices with such advanced functions have accelerated the need for network development and data analysis as well as related technological developments. Furthermore, with mobile technology providing a working environment similar to powerful computers, the development of mobile cloud technology is becoming increasingly important.

In wireless sensor networking, active research is being conducted on the interface between monitoring systems outside of greenhouses and sensor status information collection technology [13]. Instead of defining a new stack, two-way communication can be realized through protocol conversion for interfacing between the sensors and external network. This two-way communication uses existing network technology and provides easy connection to the sensor network; in addition, interfacing with the external network is accomplished through the Internet protocol (IP) or web protocol HTTP [14].

The wireless sensor network, based on IEEE 802.15.4, is standardized by IPv6 over Low-power WPAN (6LoWPAN) technology and IPv6 Routing Protocol for Low-power and Lossy Networks (RPL), which is a low-power wireless routing technology [15–18]. This wireless senor network also includes the Constrained Application Protocol (CoAP), from the IETF CoRE working group, as an application layer protocol [19, 20]. CoAP supports Representational State Transfer (REST)-based technology and sends the resource information of the sensor nodes through GET/POST/PUT/DELETE methods, which are the same as in HTTP [21, 22].

A gateway is required for data processing in order to match the two different network conditions caused by a protocol difference between the monitoring system that is connected to the external Internet and the wireless sensor network. In general, a gateway receives the requested information from the monitoring system connected to the external Internet and requests the node resource information or the information of network topology connections between nodes through the 6LoWPAN Border Router (6LBR) [23].

3   Smart Farm Environment Management System

The configuration of the ICBM-based smart farm environment management system (SFEMS) proposed by us is shown in Fig. 1. The greenhouse environment monitoring system can be divided into an environmental information collection unit, facility control unit, and the SFEMS platform based on their different functions [24].

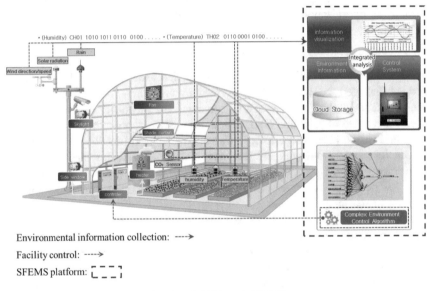

Environmental information collection: ---->
Facility control: ---->
SFEMS platform: ⌐ − − ⌐

Fig. 1. ICBM-based SFEMS

3.1   Environmental Information Collection Unit

In order to collect environmental information inside the greenhouse, the environmental information collection unit together with a facility control system measures the indoor environmental conditions (temperature, humidity, illumination) through sensors embedded in the system and stores the obtained data in a database (DB) [25, 26]. The environmental information collection unit consists of a sensor node to which all sensors are connected and a gateway that delivers the data sent from the sensor node to the DB or management server through wired or wireless connections. The sensor node and gateway are designed with the same circuit; in addition, the sensor node is designed such that it can be connected to the sensor interface through an external sensor connector. All functions of the sensor nodes can be controlled through a Micro Controller Unit (MCU); in addition, the MCU can be used for collection of environmental

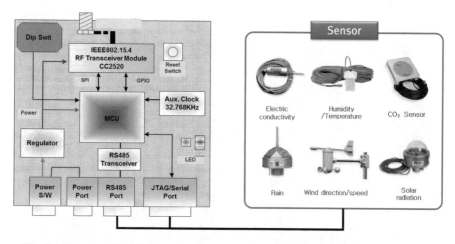

Fig. 2. Structure of the IoT sensor node used for environmental information collection

information (see Fig. 2). The information on operational status can be conveyed through the sensor node Light Emitting Diode (LED) so that users can easily identify the sensor status [27, 28].

Wireless connection units can be attached to the environmental collection unit to enable short-distance wireless communication between nodes; this can be extended over longer distances through long-distance wireless means. In this study, we designed a system using ZigBee, which is a low-cost, low power, wireless communication method; in designing the system, we consider the radiofrequency (RF) transmission characteristics and power consumption to align it with the greenhouse environmental system [29].

3.2 Facility Control Unit

As shown in Fig. 3, the facility control unit is responsible for controlling the different types of facilities to maintain the optimal conditions for crops based on the environmental information collected from various IoT sensors inside the greenhouse. The data accumulated by the environmental information collection unit are used to design the most suitable control system for the greenhouse facilities. In that vein, the facility control unit is designed to be applicable to control methods that use electric signals as well as remote control methods using communication interfaces including the Internet. When sensor data are received that are beyond the preset range for the greenhouse facilities control environment, real-time facility control can be accomplished through retrieval of the facility operation control status. By using advanced wireless sensor network technology, the unit is designed not only to enable more accurate control by analyses of the various sensing values, but also to maintain efficient control of the facilities.

The facility control unit can perform monitoring and control the growth environment, as shown in Table 1.

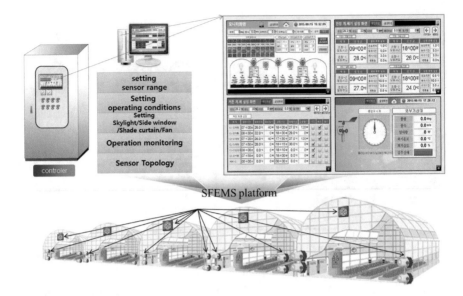

Fig. 3. Facility control unit

Table 1. Specification of greenhouse facility control service

Category	Details
Growth environment monitoring	Viewing of all sensor information
	Information on internal & external growth environment
	Monitoring of remote greenhouse status
	Verification of sensor status information
	Sensor node data
Growth environment control	Ventilation control
	Curtain control
	Fan control
Wireless system management	Sensor management
	Sensor network management

3.3   SFEMS Platform

All the information that the system obtained through real-time monitoring is immediately saved in the DB; further, the SFEMS platform is designed in a manner such that users can monitor the data through wired or wireless networks independent of time and place.

As shown in Fig. 4, the monitoring and management unit provides users with information on crop growth and the greenhouse environment; in addition, this unit also provides information from the outside weather monitoring sensors and those of facility control and management, including heating and water supply. Based on these information, the users can then manage and control the entire system. By automatically

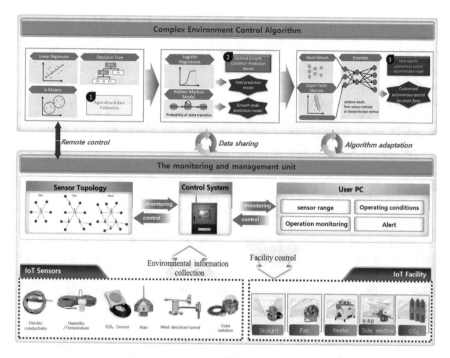

Fig. 4. SFEMS platform

providing users with information on facility control through a GUI-based display through connections with both the monitoring and the management systems of the greenhouse, the platform provides services that enable at-site as well as remote real-time monitoring and management [30].

3.4 Dual Stack Gateway

Because of the difference in network protocols between the SFEMS and wireless sensor network connected to the Internet, a dual-stack gateway is required to process the data considering the two different types of network conditions. In general, the gateway receives the requested information from the monitoring system connected to the external Internet and then requests the required resource information from the IoT nodes or topology information for the connections between nodes through the connected 6LBR [31, 32].

In order to provide information on the topology of the IoT nodes that is optimized for the greenhouse environment, we propose a bottom-up data transmission method, in which the node itself sends data to the uppermost parent category, instead of the conventional top-down search method for obtaining the tree configuration information of IoT nodes. The topology information for the wireless sensor network can be provided by visualized data routes between the nodes; in addition, a more efficient method of managing node disconnections is provided by comparing the topology information of the sensor network after a node faces a physical or link error with the previous status of the sensor. This not only enhances energy efficiency, but also helps overcome the disconnection of control signals.

Considering this, the network must be designed in a manner that enables it to define node connection and modification information within the wireless sensor network as a message that follows the CoAP protocol and delivers it asynchronously to the dual-stack gateway. This enables efficient use of wireless resources as the information will only be delivered when there is a change. Within the dual-stack gateway, sensor network topology information is structured and saved and is then sent to the cache upon the request of the SFEMS.

3.5   Network Coupling

Figure 5 shows the dual-stack gateway structure that includes the periodic request handler and cache-saving function for requesting information about the wireless sensor network. The periodic request handler receives data from the IoT nodes, which are saved in the cache after classification. Then, the gateway returns the data saved in the cache upon the request of the monitoring system or delivers it through the cache management handler when a modification to or addition of sensor nodes in the SFEMS is made.

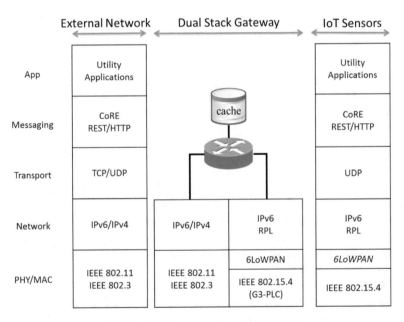

Fig. 5.   Dual-stack gateway for SFEMS

This structure enables configuration of the sensor network topology by collecting information through periodic requests and structuring the received data. Nevertheless, the collection of sensor information upon such periodic requests can result in the overloading of wireless resources by activating the system regardless of the IoT node topology modification, which makes it inefficient in terms of battery consumption. Thus, for efficient use of the limited energy resources of IoT nodes, the network must be designed to use asynchronous delivery of messages with information on connections between IoT nodes.

3.6 Protocol Design

In an IoT node that uses the CoAP protocol, the message delivered to the uppermost parent node contains the address of both the parent node and connected child node. A callback function is provided to notify the upper nodes of events including the addition and deletion of nodes; thus, the Node-List is maintained.

The called function sends the address information of the parent node and child node to 6LBR, which is the uppermost node, by referencing the Node-List. Then, 6LBR sends the requested source address and addresses of the parent and child nodes included in the Payload to the dual-stack gateway for the cache. The process of message delivery using the CoAP protocol is shown in Fig. 6. The message request uses Confirmable, which requires a response. The Confirmable message delivery in this study requires an ACK response informing the sender whether the information has been delivered normally to the uppermost parent node. When no ACK is received, the message re-sending routine is performed up to a maximum of three times.

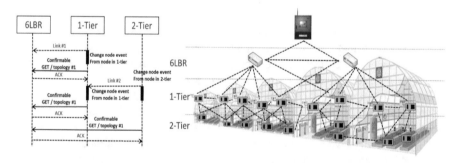

Fig. 6. Message transmission method using the CoAP protocol

4 System Implementation

The ICBM-based SFEMS proposed for this study consists of an environmental information collection unit, a facility control unit, and the SFEMS platform; it was implemented to enable user-friendly management by remotely sending the sensor information through the dual stack gateway technology.

4.1　Hardware Implementation

The SFEMS design implementation is shown in Fig. 7. In this system, we implemented the MCU control, sensor interface, and the wireless communication units. The MCU and wireless communication units were implemented on one board. The sensors that are used for collecting various greenhouse environmental data are designed so that they can be connected to the sensor interface unit as well as to the MCU control unit via the external sensor connection unit.

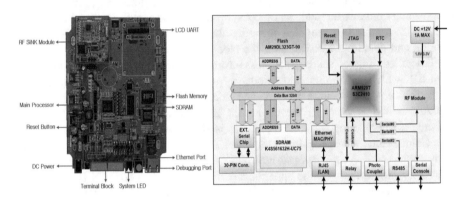

Fig. 7. SFEMS control

The MCU control unit is implemented with a low-power microprocessor, ATmega128 of Atmel Co., as the main MCU, and the sensor interface unit is implemented to enable the collection of the greenhouse environmental information through sensors for temperature, humidity, and illumination.

The facility control unit is designed to interchangeably use the IoT sensor nodes that are implemented in the environmental information collection unit as a dual stack gateway, and a single sensor node is used for the gateway.

In general, the switches and ventilation system of the greenhouse facilities are controlled through a Programmable Logic Controller (PLC), and a greenhouse facility control unit is designed to use the environmental information collected through the greenhouse environmental collection system for its operation.

4.2　Protocol Performance Evaluation

The performance comparison is shown in Fig. 8; for this, a search for all nodes was conducted (as in a periodic request) via a route from the parent to child node, which follows the same method as in a binary search. Because the search method is repeated at regular intervals, the number of wireless signals and battery consumption also increase accordingly as a function of time.

In contrast, when a node addition event occurs, because a search is performed for neighbors after these nodes are initialized, the information is sent to the uppermost parent node, and consequently, battery consumption also increases dramatically as the

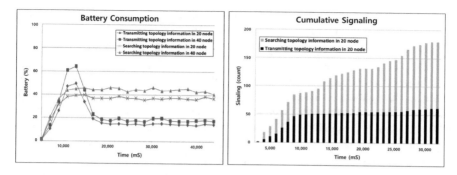

Fig. 8. Performance analysis of proposed topology

number of initial nodes increases. However, this increase in power consumption is in the beginning due to considerable use of wireless resources, thereafter the accumulation of signaling and total power consumption are constant over time.

4.3   Big Data Environmental Analysis in Smart Farm

Figure 9 shows the detailed management information on tomato growth for nine months at a tomato farm where the SFEMS was implemented [33]; it shows a comparison of the relationship between the amount of solar radiation quantity and CO_2 along with the production yield. The amount of solar radiation quantity in Weeks 12–19 was relatively higher than that in Weeks 44–51; this along with lower CO_2 concentrations led to active tomato growth. This information enables the identification

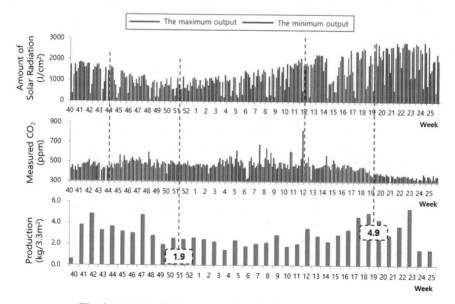

Fig. 9. Relationship between solar radiation quantity and CO_2

of the conditions when the tomato production was high; based on the obtained results, it can be confirmed that the yield was higher when the amount of solar radiation is higher, because of the relationship between the amount of solar radiation and the measured CO_2 concentration.

Thus, we can implement a consulting and automation system for growth environment management and facility control in a farm using such data analyses. Furthermore, decision making that can sustain proper growth conditions is essential in a farm because it is hard to maintain ideal growth conditions; thus, such analysis can also help elucidate the optimized conditions for specific farms. Moreover, various other data including humidity, amount of solar radiation, CO_2 concentration, amount of water supply, and nutrient solution concentration can be used to determine the appropriateness of the growth conditions.

5   Conclusions

In this study, we implemented an environmental information collection system for a greenhouse using a wireless sensor network that enables various environmental data to be obtained and saved in real-time through an environmental information collection unit, facility control unit, and SFEMS platform. Compared with the conventional node search method, this system was demonstrated to be more efficient in terms of battery usage because of its comparatively low use of wireless resources; this was confirmed through simulation and it was observed that wireless signaling did not increase as a function of time.

In a smart farm that uses our SFEMS, the most crucial technological factors include the collection of sense data, connections for interfacing, and prediction using big data analysis. Using such technology, it is possible for farms to optimize management through valuable information collection and process innovation to intelligently control the environment through prediction, and reduce costs through automation and efficiency.

Thus, when smart farm environmental control is implemented through ICBM technology, production can be enhanced through scientific analysis of environmental and growth data, consequently leading to increased income for the farmers and rural populations. Although this research verified the results of existing similar studies, we also secured our own technology. Moreover, various initiatives based on ICT can be developed to incorporate other technological areas. In summary, our proposed system can be used to efficiently and remotely monitor and control a greenhouse horticulture environment; in addition, it can be extended to a smart factory or smart home.

Acknowledgments. This research was supported by the MSIP (Ministry of Science, ICT and Future Planning), Korea, under the ITRC (Information Technology Research Center) support program (IITP-2018-2013-1-00877) supervised by the IITP (Institute for Information & communications Technology Promotion).

References

1. Lee, M., Hwang, J., Yoe, H.: Agricultural production system based on IOT. In: 16th International Conference on Computational Science and Engineering (CSE), 2013 IEEE, pp. 833–837. IEEE (2013)
2. Lee, M., Yoe, H.: Analysis of environmental stress factors using an artificial growth system and plant fitness optimization. Biomed. Res. Int. **2015**, 6 (2015)
3. Jayaraman, P., Yavari, A., Georgakopoulos, D., Morshed, A., Zaslavsky, A.: Internet of Things platform for smart farming: experiences and lessons learnt. Sensors **16**, 1884 (2016)
4. Weber, R.H., Weber, R.: Internet of Things. Springer, Heidelberg (2010)
5. Wu, G., Talwar, S., Johnsson, K., Himayat, N., Johnson, K.D.: M2 M: from mobile to embedded internet. IEEE Commun. Mag. **49**, 36–43 (2011)
6. Lynggaard, P., Skouby, K.: Complex IoT systems as enablers for smart homes in a smart city vision. Sensors **16**, 1840 (2016)
7. Zhang, Q., Cheng, L., Boutaba, R.: Cloud computing: state-of-the-art and research challenges. J. Internet Serv. Appl. **1**, 7–18 (2010)
8. Mell, P., Grance, T.: The NIST Definition of Cloud Computing (2011)
9. Buyya, R., Yeo, C.S., Venugopal, S., Broberg, J., Brandic, I.: Cloud computing and emerging IT platforms: Vision, hype, and reality for delivering computing as the 5th utility. Future Gener. Comput. Syst. **25**, 599–616 (2009)
10. Manyika, J., Chui, M., Brown, B., Bughin, J., Dobbs, R., Roxburgh, C., Byers, A.H.: Big Data: The Next Frontier for Innovation, Competition, and Productivity (2011)
11. Chen, M., Mao, S., Zhang, Y., Leung, V.C.: Big Data: Related Technologies, Challenges and Future Prospects. Springer, New York (2014)
12. Jara, A.J., Zamora, M.A., Skarmeta, A.F.: An Initial Approach to Support Mobility in Hospital Wireless Sensor Networks Based on 6LoWPAN (HWSN6) (2010)
13. Kuladinithi, K., Bergmann, O., Pötsch, T., Becker, M., Görg, C.: Implementation of CoAP and its application in transport logistics. In: Proceedings of IP + SN, Chicago, IL, USA (2011)
14. Fielding, R., Gettys, J., Mogul, J., Frystyk, H., Masinter, L., Leach, P., Berners-Lee, T.: Hypertext Transfer Protocol–HTTP/1.1 (1999)
15. Mulligan, G.: The 6LoWPAN architecture. In: Proceedings of the 4th Workshop on Embedded Networked Sensors, pp. 78–82. ACM (2007)
16. Shelby, Z., Bormann, C.: 6LoWPAN: The Wireless Embedded Internet. John Wiley & Sons, Chichester (2011)
17. Winter, T.: RPL: IPv6 Routing Protocol for Low-Power and Lossy Networks (2012)
18. Hui, J.W.: The Routing Protocol for Low-Power and Lossy Networks (RPL) Option for Carrying RPL Information in Data-Plane Datagrams (2012)
19. Bormann, C., Castellani, A.P., Shelby, Z.: CoAP: an application protocol for billions of tiny internet nodes. IEEE Internet Comput. **16**, 62 (2012)
20. Kovatsch, M., Duquennoy, S., Dunkels, A.: A low-power CoAP for Contiki. In: 2011 IEEE Eighth International Conference on Mobile Ad-Hoc and Sensor Systems, pp. 855–860. IEEE (2011)
21. Raza, S., Trabalza, D., Voigt, T.: 6LoWPAN compressed DTLS for CoAP. In: 2012 IEEE 8th International Conference on Distributed Computing in Sensor Systems, pp. 287–289. IEEE (2012)
22. Ma, X., Luo, W.: The analysis of 6LoWPAN technology. In: 2008 IEEE Pacific-Asia Workshop on Computational Intelligence and Industrial Application (2008)

23. Efendi, A.M., Negara, A.F.P., Kyo, O.S., Choi, D.: A design of 6LoWPAN routing protocol border router with multi-uplink interface: ethernet and Wi-Fi. Adv. Sci. Lett. **20**, 56–60 (2014)
24. Lee, M., Kim, H., Yoe, H.: Intelligent environment management system for controlled horticulture. In: 2017 4th NAFOSTED Conference on Information and Computer Science, pp. 116–119 (2017)
25. Lee, M.-h., Eom, K.-b., Kang, H.-j., Shin, C.-s., Yoe, H.: Design and implementation of wireless sensor network for ubiquitous glass houses. In: Seventh IEEE/ACIS International Conference on Computer and Information Science 2008, ICIS 08, pp. 397–400. IEEE (2008)
26. Han, J., Haihong, E., Le, G., Du, J.: Survey on NoSQL database. In: 6th International Conference on Pervasive Computing and Applications (ICPCA) 2011, pp. 363–366. IEEE (2011)
27. Lee, M.-h., Yoe, H.: Comparative analysis and design of wired and wireless integrated networks for wireless sensor networks. In: 5th ACIS International Conference on Software Engineering Research, Management & Applications (SERA 2007), pp. 518–522. IEEE (2007)
28. Gardner, J.W., Varadan, V.K., Awadelkarim, O.O.: Microsensors, MEMS, and Smart Devices. Wiley, New York (2001)
29. Capella, J., Campelo, J., Bonastre, A., Ors, R.: A reference model for monitoring IoT WSN-based applications. Sensors **16**, 1816 (2016)
30. Zikopoulos, P., Eaton, C.: Understanding Big Data: Analytics for Enterprise Class Hadoop and Streaming Data. McGraw-Hill Osborne Media, New York (2011)
31. Shen, W., Xu, Y., Xie, D., Zhang, T., Johansson, A.: Smart border routers for ehealthcare wireless sensor networks. In: 7th International Conference on Wireless Communications, Networking and Mobile Computing (WiCOM) 2011, pp. 1–4. IEEE (2011)
32. Chang, H.-L., Wang, C.-G., Wu, M.-T., Tsai, M.-H., Lin, C.-Y.: Gateway-assisted retransmission for lightweight and reliable IoT communications. Sensors **16**, 1560 (2016)
33. Choe, Y.C.: Analysis method of measurement data for solution of difficulties in agricultural field. Report, Seoul National University (2015)

Design of I-beacon Based Distribution Route Management System to Prevent the Spread of Livestock Disease

Hojin Yang and Hyun Yoe[✉]

Department of Information and Communication Engineering,
Sunchon National University, Suncheon,
Jeollanam-do 57922, Republic of Korea
ghwlsdl27@naver.com, yhyun@scnu.ac.kr

Abstract. Livestock diseases such as foot-and-mouth disease and AI have continued to spread and the damage of domestic livestock farmers is increasing. This leads to a national loss due to infectious proliferation as well as an economic loss due to productivity and income decrease in livestock farmers, which has a serious adverse effect on the development of livestock industry. Representative examples of such livestock diseases include foot-and-mouth disease and AI virus. A major source of infection of foot-and-mouth disease and AI spread is virus transmission by livestock transportation vehicles. To solve this problem, an I-beacon-based distribution management system was designed to prevent the spread of livestock diseases caused by livestock transportation vehicles. I-beacon's Proximity UUID, Major and Minor identifier data are used to determine whether the GPS and Bluetooth communication technology is used by identifying the indoor/outdoor location of the farm, and the location information data of the transportation vehicle is identified, Provides real-time monitoring service by visualizing location information and distribution route data by grasping information and distribution route. When the system design is used to detect livestock diseases, it is possible to prepare for the proliferation of livestock diseases by investigating the spread of livestock diseases based on the distribution route record, focusing on the point where the first livestock disease occurs.

Keywords: I-beacon · Livestock disease · Proximity UUID
Management system

1 Introduction

Livestock disease not only leads to economic losses due to productivity and income decline in livestock farmers, but also leads to national loss due to infectious spread [1, 6]. The onset of livestock diseases such as foot-and-mouth disease and AI (avian influenza) continues to increase the damage of domestic livestock farmers. Considering that the total amount of domestic livestock industry production is 11 trillion won, the amount of damages caused by livestock diseases and related industries is 2.2 trillion annually, which means that livestock diseases have a serious impact on the development

© Springer Nature Switzerland AG 2019
R. Lee (Ed.): SNPD 2018, SCI 790, pp. 57–65, 2019.
https://doi.org/10.1007/978-3-319-98367-7_5

of livestock industry [3]. As a representative example of such livestock diseases, there are foot-and-mouth disease and AI virus. The rapid spread of domestic animal diseases such as foot-and-mouth disease and AI has caused enormous damage to the domestic animals. The second reason is that domestic livestock industry has evolved into a factory type dense livestock farm, and it has evolved into a structure where virus propagation between farms can be easily done [1, 3]. In addition, there are various problems such as limit blockage prevention at the base area and inadequate blockage prevention at farmhouse level [2]. Generally, it is not only direct contact with the feces of livestock infected with virus, but also propagation through people, vehicles, etc. [3, 5]. In the case of AI, the 212 major routes of AI infection that occurred in 2013/2014 were 28.3% of wild birds, 27.4% of house owners, 26.9% of vehicles [4], and 83 cases of foot and mouth disease major outbreak in foot-and- Compared with 54.2% for livestock transport vehicles, 18.9% for feed vehicles, and 12.6% for nearby radio waves, all of which have a very high risk of livestock disease caused by vehicles [5].

In order to solve this problem, it is necessary to study the system to prevent spread of livestock diseases in order to prevent the spread of livestock diseases by livestock transportation vehicles. In the case of livestock diseases, if the early response time shortening is a way to reduce the economic loss, the economic loss can be reduced if the livestock transport vehicle can respond to the spread of livestock disease early [6]. However, Korea has a characteristic that it is difficult to prevent the spread of livestock disease in the early stage because the spread of the livestock disease spreads rapidly due to the development of traffic compared to the relatively small land area in Korea. However, if the distribution route of the livestock transportation vehicle can be grasped, it is possible to prevent the spread of livestock disease by minimizing the damage caused by the livestock disease by focusing on the distribution route based on the point where the livestock disease has occurred Can be. In this paper, we try to design an I-beacon-based distribution route management system to prevent the spread of livestock diseases by preliminarily grasping the route of infection to livestock transport vehicles. In this paper, we discuss related research in Sect. 2, and describe the I-beacon-based livestock distribution channel system proposed in Sect. 3. Finally, Sect. 4 presents conclusions.

2   Related Research

I-beacon, a technology standard developed by Apple at the World Wide Developer Conference (WWDC) in 2013, is a low-power, short-range communication technology using Bluetooth low energy (BLE). This is a transmitter that periodically broadcasts beacon signals and devices that support Bluetooth version 4.0 or later receive signals and calculate distances to measure their position [7, 11]. I-beacon can be used for all devices with BLE chips such as BLE-USB-Sticks or Arduino-Boards as well as for Apple-related products. It is based on BLE communication. BLE communication procedure using I- And the scanning operation [8].

① Advertising: Procedures and procedures for requesting connection through Broadcasting operation in I-beacon device and transmitting advertisement packets at regular intervals

② Scanning: Procedure and process of receiving the advertisement packet transmitted by I-beacon device

To use the I-beacon service, the user's smart device must be iOS 7.0 or Android 4.3 or later. If the condition is satisfied, the smart device's Bluetooth mode must be activated [10]. If the service application is installed on the user's smart device It is possible to use. If these conditions are satisfied, the beacon signal can be transmitted and received via the smart device. To use the beacon service, the following four factors are needed [9].

① beacon device SDK for smart devices
② Bluetooth beacon device hardware
③ Service application using beacon device SDK
④ beacon-based service platform

2.1 Data Structure of I-beacon

The structure of the data transmitted through I-beacon is shown in Fig. 1. The data structure area of I-beacon used in this paper is 'PDU' area. PDU is composed of minimum 2 bytes ∼ maximum 39 bytes. Of the PDU area used in I-beacon, 31 bytes is used for data [12]. The system processes the beacon recognition using the Major, Minor, and Proximity UUIDs of PDU data. The data structure of PDU area is shown in Fig. 2.

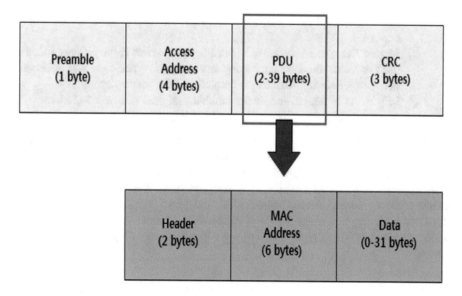

Fig. 1. I-beacon data structure

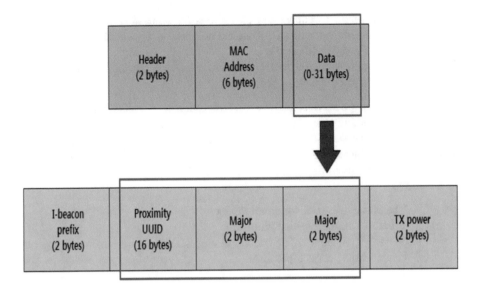

Fig. 2. Configuring PDU area data

Proximity UUID, Major and Minor act as identifiers of I-beacon, continuously transmitting beacon signals, [12] and acting as interactions between I-beacon and smart devices through the configuration of proximity UUID, Major and Minor values.

① Proximity UUID: The Proximity UUID is a standard identification system that generates a unique identification number for the beacon. The purpose of ID is to provide basic information of I-beacon by distinguishing I-beacon in network and all other signals of network that user's smart device can not control.

② Major: The number assigned to the I-beacon, which is the primary identifier used to locate the I-beacon, using any changes to fine-tune and distinguish between where the I-beacon is located and the service group.

③ Minor: As in Major, this is the number assigned to I-beacon, which is the secondary identifier used to locate the I-beacon.

I-beacon conveys location information via Proximity UUID, Major and Minor values and facilitates management of I-beacon. In this study, I-beacon signal is transmitted/received by using proximity UUID, major and minor identifiers, and it acts as an interaction between I-beacon and smart device. When a livestock transport vehicle with an I-beacon device based on the transmitted and received I-beacon signal enters and exits the area of the I-beacon transmitter and receiver, the user can monitor the location and the distribution route of the vehicle in real- To provide a notification service.

3 System Design

3.1 I-beacon-based Distribution System Overview

The overall process of the I-beacon-based distribution management system for preventing the spread of livestock diseases proposed in this paper is shown in Fig. 3.

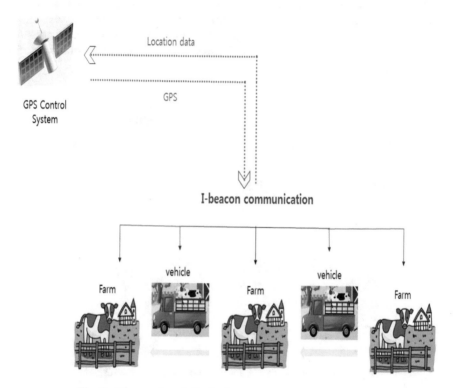

Fig. 3. I-beacon-based distribution management system process

The livestock transit vehicle has a built-in PDA capable of receiving Bluetooth and GPS signals and attaches an I-beacon device. The ID of the I-beacon device is an identifier for the livestock transport vehicle and provides information about the corresponding vehicle in the distribution process. The I-beacon device is an identifier for the corresponding vehicle, and accesses the farm where the I-beacon transmitter and receiver. The ID of the vehicle acquired based on the data obtained from the recognition of the I-beacon device and the location and the distribution route of the vehicle received from the GPS are identified. The location and distribution route data of the livestock transportation or the transportation of the livestock according to the route are transmitted to the server, the I-beacon information and the location information data which have been subjected to the authentication process of the server are transmitted to the database server, the database server sends the data requested by the application. The location information and distribution route data of the vehicle stored in the database server provides real-time monitoring service by visualizing information so that the recognition processing through the application based on the server and the location and the distribution route of the transportation vehicle can be grasped to the user.

3.2 I-beacon Based Distribution Management System Design

Elements that make up this system are ① smart devices. An ② I-beacon transmitter installed on the farm. ③ I-beacon devices attached to livestock transport vehicles, ④ servers, ⑤ database servers for storing and managing information on databases and vehicles, and ⑥ mobile application and ⑦ users. Figure 4 shows the structure of the whole system.

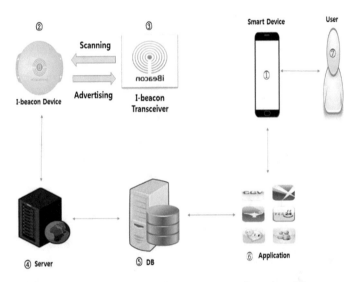

Fig. 4. Distribution management system configuration diagram

As shown in Fig. 4, the I-beacon receiver and the I-beacon device are installed in each farm, and the I-beacon device is attached to the I-beacon receiver. -beacon When entering the signal area of the receiver, it performs the location based service and receives the signal of I-beacon every time it approaches the farm and proceeds with the scanning step. The i-beacon device scanning step proceeds as shown in Fig. 5.

At this stage, information on each farm is identified to determine the indoor and outdoor locations of the farm. The I-beacon message contains the Major and Minor values as shown in Fig. 2. If the I-beacon signal is received, the minor value of the I-beacon signal, as in 5, identifies the indoor or outdoor location of the farm. If the Minor value is odd, it distinguishes the farm location indoors, reduces battery consumption by not using the GPS function, and when the Minor value is even, it separates the farm location outdoors and activates the GPS function, and receives location information. As shown in Fig. 2, I-beacon message is used to scan the data of the farm through the Major value, and the location and distribution route data of the farm can be grasped. The

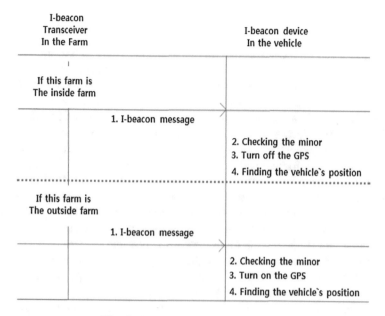

Fig. 5. I-beacon device scanning step

location data of the scanned farm is authenticated through the server before it is stored in the database. When the I-beacon device of the transportation vehicle is scanned in the I-beacon transmitter and receiver, the I-beacon Proximity UUID data including the vehicle identification code such as the IP address of the vehicle is input to the I- From the identification code information of the vehicle, the position of the vehicle and the distribution route data are transmitted to the server, and the server requests a vehicle authentication and authentication request message (Minor, MAC device).

In response to this, the server processes the Minor value and the MAC device by performing a hash function and processes the response result of the authentication of the vehicle information based on the message. The result of the authentication response is sent to the I-beacon. The position and route data of the vehicle can be grasped through the Proximity UUID of the transportation vehicle. The acquired data is stored in the database server, UUID is a structure that can obtain records of location information and distribution route data. Proximity UUID is required to access data. When data request is received from server, all data stored in database server can be transmitted.

When the stored data accesses the I-beacon device attached to the livestock transportation vehicle through the application to the user's smart device, the I-beacon device continuously transmits the advertising packet, and receives the packet through the smart device of the user. Proximity UUID After checking the process, the smart device executes the corresponding application and transmits the user's information to the server for authentication. When the authentication is completed, the server inquires the user about the location information and route information of the livestock transportation vehicle obtained from the I-beacon device and provides the information of the corresponding vehicle in real time.

4 Conclusion

In this paper, I-beacon based distribution management system to prevent the spread of livestock disease has designed a distribution management system for livestock transportation vehicles based on I-beacon technology to prevent the spread of animal diseases such as foot-and-mouth disease and AI. In this system, I-beacon's proximity UUID, major and minor identifier data are used. Major and minor identifiers are used according to actual location of the farm, and whether or not GPS and Bluetooth communication technology are used. By utilizing the proximity UUID, it is possible to identify the location information data of the transport vehicle with I-beacon device and to design the location information and distribution route of the livestock transport vehicle. Through this system design, the user can monitor information about the location and distribution route of livestock transportation vehicles in real time, and based on the point where livestock disease occurs, the route of livestock disease can be inferred based on the distribution route record It is possible to investigate farms based on distribution route records, to investigate the spread of livestock diseases on suspected farmers, and to take action on disease-spread farms to prevent the spread of livestock diseases. In addition, it is anticipated that it will be possible to minimize damage to the farmhouse in response to the spread of livestock disease. Future research will focus on cloud-based distribution management systems. By using cloud computing technology, it is expected that the initial introduction cost of the distribution management system can be lowered, and the distribution management system can be distributed more smoothly to farmers as it is installed as a mobile-based service.

Acknowledgements. This research was supported by the MSIP (Ministry of Science, ICT and Future Planning), Korea, under the ITRC (Information Technology Research Center) support program (IITP-2018-2013-1-00877) supervised by the IITP (Institute for Information & communications Technology Promotion).

References

1. Lee, S.h.: AI (bird influenza), spread of foot and mouth disease Issues and Challenges. Issue Anal. **272** (2017)
2. Kim, H.-g., Yang, C.-j., Yoe, H.: Design and implementation of livestock disease forecasting system. Korean Inst. Commun. Sci. **37**(12), 1263–1270 (2012)
3. Lee, M.h.: Design of animal disease monitoring system using ITtechnology. In: Science and Technology Policy, pp. 3–16(14) (2011)
4. Lee, J., Yoe, H.: Clustering and classification to characterize daily electricity demand. In: Korea Telecom Society Conference, pp. 89–90(2) (2017)
5. Kim, J.S.: Environmental problems and civic science due to foot and mouth disease policy failure. In: Environmental Sociology Research ECO, pp. 85–119 (2011)
6. Kang, Y.-J., Choi Lee, D.-O.: Development a animal bio-information monitoring device. J. Korea Entertain. Ind. **6**(2), 101–106 (2012)
7. Kohne, M., Sieck, J.: Location-based services with iBeacon technology. In: Second International Conference on Artificial Intelligence, Modelling and Simulation, pp. 18–20 (2007)

8. Lee, J.-w., Im, S.-y., Kim, D.s., Shin, S.-H., Roh, B.-h.: Analysis of BLE structure and evaluation of scanning. In: SpeedKorea Telecom Society Winter Synthesis Conference, pp. 343–344 (2016)

9. Han, S.-B., Song, J.-H.: A push technique of the product informations by Beacon's MacAddress. J. Korean Soc. Compos. Mater. 13–21 (2016)

10. Joo, J., Lee, H., Kim, J., Han, D.S.: A dual beaconing scheme for effective context awareness in vehicular ad hoc networks. J. Korean Inst. Commun. Sci. **39**(2), 114–122 (2014)

11. Jung, H.-H., Shin, H.-H., Nam, C.-S., Shin, D.-R.: Design of the automatic access authentication system using iBeacon. In: Korea Computer Information Association, pp. 217–218 (2015)

12. Kim, D., Kim, S., Jin, S.: The research on iBeacon technology trend and issue. In: Korea Computer Convention, pp. 390–392 (2014)

Optimal Energy Management System Design for Smart Cattle Shed

Hyeono Choe and Hyun Yoe[✉]

Department of Information and Communication Engineering,
Sunchon National University, Sunchon, Jeollanam-do 57922, Republic of Korea
romantic_rad@naver.com, yhyun@scnu.ac.kr

Abstract. As the importance of energy consumption has recently increased, even Cattle shed facilities are using a lot of energy to grow livestock. However, as the energy consumption increases, wasted energy is generated, which is a big problem. To solve these problems, we designed the optimal energy management system using ICBM technology. Data from IoT sensors and controllers are collected within smart cattle shed and classified through analysis of Big Data in the database. The classification of data is from cloud systems to derive optimal energy management methods through SVR algorithms. And displays the results to the user on the mobile, allowing them to monitor the power value and control the device via the mobile. Through the designed system, the energy consumption of existing smart cattle shed compared to a barn applied with an optimal energy management system, there is not much difference when the usage is high. When the usage is low, it is set to use only the optimum energy and there are many differences. By analyzing the information of the surrounding environment through the design, it is possible to manage the optimum energy anytime and anywhere in real time, by providing power usage guidelines for smart cattle shed, energy usage and energy costs (power charges) can be saved.

Keywords: Smart cattle shed · Energy management · Big data
Cloud · IoT

1 Introduction

In recent years, the importance of energy consumption has increased in Korea, in order to grow livestock in stables, energy management system is becoming an important core technology because it occupies a large portion of energy consumption due to air-conditioning and IoT equipment. However, unlike expectations, the reason why optimal energy management systems are not utilized not only is there not enough objective data to prove about the energy savings that can best account for the importance of an optimal energy management system, this is because the system for managing optimum energy is not spreading [1, 2]. In order to manage optimal energy for smart cattle shed energy reduction is necessary first. To save energy, it is divided into architectural and systemic components. Architectural factors include appearance and location of the house, solar radiation, and insulation performance improvement. Systematic factors include system efficiency improvement, sensor efficiency improvement, control

© Springer Nature Switzerland AG 2019
R. Lee (Ed.): SNPD 2018, SCI 790, pp. 66–75, 2019.
https://doi.org/10.1007/978-3-319-98367-7_6

method, and alternative energy use [3, 4]. Among these, we decided to save energy in the heating and cooling system and power system used in the cattle shed, this paper aims to design energy management system for smart cattle shed by searching energy saving method in stables and using efficient control method.

2 Smart Cattle Shed Summary

Smart cattle shed is a system that integrates network (internet) and automation technology in existing cattle shed, so that it can reduce the labor force by applying the livestock production and distribution to the barn, Putting improvements and productivity of livestock breeding environment with emphasis.

In Smart Cattle shed, it operates through environmental management, specification management, and management. First, the environmental management can collect information and monitor the inside of the cattle shed (temperature, humidity, power outage, fire), outside (temperature, humidity, wind direction, wind speed), and CCTV. The second specification management is carried out through the control of the feed bin manager, the Auto Sorting, the automatic feeder, and the water controller. The third management establishes and analyzes the management plan through production management, business administration, and auto sorting management [5] (Fig. 1).

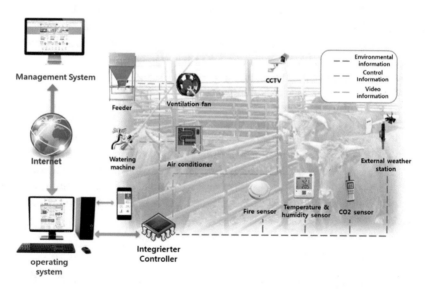

Fig. 1. Smart cattle shed system configuration diagram

3 Related Research

In the related research, it is examined whether optimal energy operation system is applied to ICBM technology for smart cattle shed optimal energy management.

3.1   IoT and Sensor Technology

IoT technology is a fusion of Internet (middleware), hardware (sensor) and information. In other words, IoT technology is a technology that uses sensors to realize communication through the Internet for everything surrounding the environment. Typically, the smart meter is a developed power usage measuring device that is applied to such IoT devices. Specifically, smart meters are installed in distribution boards and switchboards to accurately measure power usage [6, 7]. A digital meter is also a power meter. The difference is that smart meters monitor not only power usage but also electricity rates, while digital meters are not.

3.2   Cloud Technology

Cloud technology has emerged as a global issue, addressing the traditional IT environment of the computing environment. Cloud computing is a computing paradigm that enables remote computing resources to be remotely accessed and utilized over a network. In other words, when a user uses a program or data stored in a PC or various terminal-type devices themselves, or stored programs, data, etc. in a large computer, this is a user-oriented computing concept that allows users to perform necessary tasks remotely over a network without having to worry about time and place.

With the advancement of cloud technology, especially in the energy sector, consumers are able to transmit their energy consumption status to a server anytime and anywhere, and transmitted remotely, and these data are analyzed on the server. Analyzed data allows consumers to see energy consumption patterns anywhere, anytime. If the energy usage exceeds the average, an alarm function is provided to encourage users to voluntarily save energy. Considering that about 20–30% of electric power is used as an unnecessary waste, cloud technology has an advantage of providing energy consumption and pattern information to consumers in real time, thereby enabling users to know and conserve energy themselves [8, 9]. Indeed, research has shown that users are aware of energy usage information and have 10–20% energy savings when they are frequently informed about energy usage. In addition, it can provide high-level energy use pattern analysis service by comparing and analyzing energy use information gathered in various buildings and residences. In summary, cloud technology collects usage information and provides processed data like a data warehouse anytime, anywhere, in real time.

3.3   Big Data Technology

If cloud technology serves as a data warehouse to store and collect data, Big Data Technology is a technology that analyzes storage devices such as OLAP. With the recent advancement of SNS technology, the advent of the IoT era through the development of network technology has led to the production of comparable data both quantitatively and qualitatively. In particular, the increasing use of SNS users has led to a large portion of the existing informal data on the server. Therefore, it is necessary to develop a technology that can generate new value by analyzing such data. This technology is big data. Big data is not simply a large amount of data retention technology that cannot be compared with existing ones, but rather a technique for processing and analyzing informal data [10]. Consumer energy use pattern analysis, Energy optimization due to efficient use of building facilities, Energy consumption due to external environment (temperature, humidity, illumination), Accurate multidimensional analysis is required from a large number of meter data in a building in order to derive the results such as optimized energy use depending on the environment (irradiation, temperature) depending on the location of the building. In particular, the amount of data is considerable. Even though it is a five-story building, the amount of data accumulated in one hour is more than tens of thousands and occupies more than 100 MB. Furthermore, a fast engine and a highly accurate system are required to provide real-time analysis and presentation to consumers. It is especially suitable for automation such as AI, especially for optimization systems using big data engines [11].

3.4 Mobile Technology

As mobile technology has developed, functions such as computers have become possible for mobile. Mobile is a term referring to smartphones and tablet PCs. Recently, Smart Watch and Google Glass, which will be developed in the future, are also in the mobile category. Beyond the functions of the PC, mobile technology enables interchange of data reception anytime and anywhere, providing services such as GPS information, analysis of SNS contents, and health information in real time. The emergence of smart phones with various functions has led to the development of network development and data analysis and the development of technology. Especially, since mobile technology provides work environments such as PC, development of mobile cloud technology like mobile devices is beginning to become important [12].

As mentioned earlier, the cloud technology is applied to mobile technology to provide energy analysis system to consumers in real time. Remotely control buildings remotely with mobile technology. In other words, even if the building manager is out of the building, it is possible to manage the situation information of the buildings in real time and remotely manage it with the mobile one, so that the energy waste can be minimized. In addition, the alarm function, setting the maximum energy usage step by step, to allow the management of the mobile beyond the energy use exceeding standards, making it easier to manage energy efficiency.

4 Design

We will design an optimal energy management system for smart cattle shed by related studies. Figure 2 shows a smart cattle shed optimal energy management system design using ICBM.

As shown in Fig. 2, the data of the IoT sensor and the controller are collected in the database, we derive optimal energy efficiency method through SVR algorithm in cloud system by using data classified by big data analysis in database. And displays the results to the user on the mobile, allowing them to monitor the power value and control the device via the mobile. The following figure shows the system flow chart.

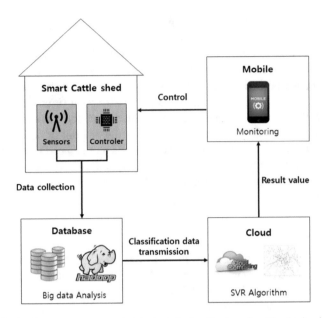

Fig. 2. Optimal energy management system design for smart cattle shed using ICBM

4.1 Data Collection and Storage

Big data analysis of power usage using Hadoop in the database. First, a big data analysis server is configured, and power data for each node is distributed and stored. We then analyze the data by applying data mining techniques. Figure 3 illustrates the process of collecting power data.

The steps of data collection are divided into DP server (Data Portal Server) collection, migration and loading. The DP server stores the power data measured by each gateway.

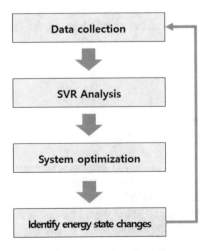

Fig. 3. System data flow chart

The collected data is transferred from Hadoop to HBase of Hadoop system for analysis of big data as shown in Fig. 3 above. Sqoop is used for data movement. The goal of the data collection phase is to efficiently collect data on power usage from each gateway, transfer it reliably to a pre-built Hadoop system, and prepare for analysis (Fig. 4).

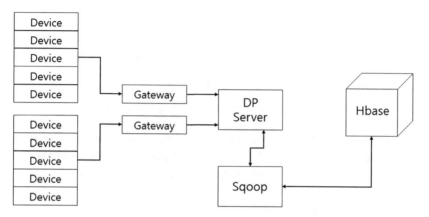

Fig. 4. Sensor and controller power data flow

4.2 Data Analysis

Based on the data classified in the database, SVR algorithm is applied in the cloud to derive optimal energy usage. First, SVR is derived from computing a linear regression function in a high dimensional space. Given a set of data belonging to one of two categories, the SVR algorithm builds a non-stochastic binary linear classification model based on a given set of data to determine which category the new data belongs to. SVR can be used to obtain high level of prediction accuracy through SVR if the main parameters of electric energy use and equipment performance are selected appropriately. The three single algorithm analysis models described above were developed to overcome the inherent limitations of a single analytical method and improve predictive performance when applied in sequential or parallel combination (Fig. 5).

SVR algorithm is used to predict smart cattle shed energy, Provides guidelines for using optimal energy based on predicted cattle shed energy. Based on the predicted data, electricity usage can be reduced through equal control. According to the electricity rate calculation method Monthly Electricity Charge = Base Rate + Usage Charge = [7430 * (peak year)] + Usage charge per time zone. For example, in the case of cattle shed, the annual peak value of 1437 kW -> 1238 kW would be 10,676,000 won -> 9,198,000 won, saving 1,478,000 won per month and the annual savings would be about 17,743,000 won, a 2.9% decrease.

4.3 Optimize System Control

Optimal control of sensor and controller (Temperature, humidity, irrigation, air-conditioner, ventilator, CO2, external weather station, etc.) usage within the smart

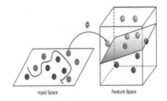

Support Vector Regression

- energy consumption and system
  performance prediction
- High prediction accuracy

Fig. 5. SVR (Support Vector Regression) algorithm

cattle shed by optimized energy through the cloud. In addition, mobile enables users to easily monitor real-time analysis data remotely and control facilities. With this control, users can receive information about energy and receive user feedback.

4.4   Optimize System Control

The result of using the energy amount and the optimum energy used in the existing Smart cattle shed is shown as the following figure by applying the smart cattle shed optimal energy management system using ICBM.

In Fig. 6, the blue line represents the existing smart cattle shed energy usage and the red line represents the optimal energy usage. When the amount is high, it does not

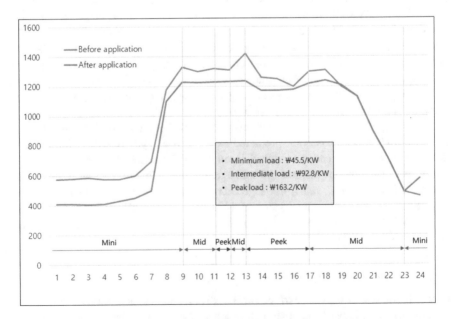

Fig. 6. Simulation result of applying optimal energy management system

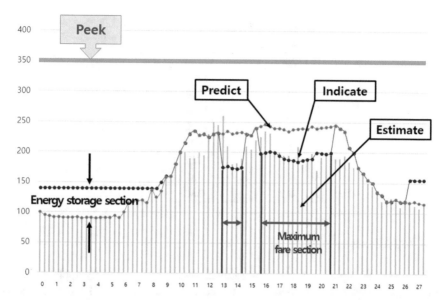

Fig. 7. Optimal energy management system Rate forecast result by electricity usage value

show much difference, but when the usage is low, it is set to use only the optimum energy, and it is seen that there are many differences. Figure 7 shows the charges based on the electricity consumption. Actual usage was measured similar to the predicted value through the SVR algorithm (Table 1).

Table 1. Cost data table by electricity usage

	Before application	After application
1	575	410
2	580	412
3	585	409
4	578	410
5	575	430
6	600	450
7	695	500
8	1180	1100
9	1330	1230
10	1300	1225
11	1320	1230
12	1310	1233
13	1423	1235
14	1260	1170
15	1250	1173

(*continued*)

Table 1. (*continued*)

	Before application	After application
16	1200	1180
17	1300	1220
18	1310	1240
19	1200	1210
20	1130	1130
21	900	897
22	715	712
23	490	488
24	462	580

5 Conclusion

In this paper, the optimal energy management system is designed by applying ICBM technology to smart cattle shed. By analyzing the information of the surrounding environment through the design, it is possible to manage the optimum energy anytime and anywhere in real time, energy usage and energy costs (power charges) can be saved by providing power usage guidelines for Smart cattle shed.

Future researches are becoming more diverse with their awareness of energy consumption. However, the same energy analysis methodology cannot be applied everywhere, we think that a solution should be provided based on data such as the number of livestock and the size of the house. Therefore, it is necessary to investigate various power consumption prediction by applying SVR algorithm, ANN and other algorithms. In the future, we will build an optimal energy management system and plan to conduct further research based on power data and charge data. It also improves convenience by increasing data delivery speed using a single platform, from data collection to storage in the cloud.

Acknowledgements. This research was supported by the MSIP (Ministry of Science, ICT and Future Planning), Korea, under the ITRC (Information Technology Research Center) support program (IITP-2018-2013-1-00877) supervised by the IITP (Institute for Information & communications Technology Promotion).

References

1. Song, J.-Y., Kim, J.: A Study on Effectiveness Analysis with Application and Operation of Active Building Energy Management System, pp. 24–27. The Society of Air-Conditioning and Refrigerating Engineers of Korea (2015)
2. Cho, S., Lee, S.-B.: Comparing methodology of building energy analysis - comparative analysis from steady-state simulation to data-driven analysis. KIEAE J. **17**(5), 77–86 (2017)
3. Hwang, J.-W., Ahn, B.-C.: Experimental study on optimal operation strategies for energy saving in building central cooling system. Korea Acad. Ind. Coop. Soc. **14**(9), 4610–4615 (2013)

4. Park, D., Yoon, S.: Clustering and classification to characterize daily electricity demand. J. Korean Data Inf. Sci. Soc. **28**(2), 395–406 (2017)
5. Chang, K., Im, Y., Yang, J., Yoon, S.: Supply and the optimal management system of energy in smart farm. Korean Soc. New Renew. Energy **2017**(5), 145 (2017)
6. Youk, C., Lim, H., Lee, J.: A proposal of an energy efficiency management system using Big data technology. Korea Comput. Congr. **2014**(6), 1997–1999 (2014)
7. Basak, D., Pal, S., Patranabis, D.C.: Support vector regression. Neural Inf. Process. **11**, 23–24 (2007)
8. Oh, M.: Trends in domestic and international standardization of cloud computing. ETREND **29**(04), 59–71 (2014)
9. Sarah, D.: The effectiveness of feed back on energy consumption. A review of DEFRA of the literature on Metering, Billing and direct Displays. Environ. Change Inst. **9**, 1–21 (2006)
10. Farqui, A., Sergici, S., Sharif, A.: The impact of informational feedback on energy consumption: a survey of the experimental evidence. Energy J. **35**, 1598–1608 (2009)
11. Huang, Y.-F., Chang, S.-H.: Mining optimum models of generating solar power based on big data analysis. Sol. Energy **155**, 224–232 (2017)
12. Bajad, R.A., Sinhal, A.: Cloud computing on smartphone. Comput. Eng. Intell. Syst. **4**(3), 73–78 (2013)

A New Polynomial Algorithm for Cluster Deletion Problem

Sabrine Malek[1(✉)] and Wady Naanaa[2]

[1] Faculty of Economics and Management of Sfax, Tunis, Tunisia
sabrine.malek@gmail.com
[2] National Engineering School of Tunis, Tunis, Tunisia
wady.naanaa@gmail.com

Abstract. Given a simple graph G, the cluster deletion problem asks for transforming G into a disjoint union of cliques by removing as few edges as possible. This optimization problem is known as the *Cluster Deletion* (CD) problem and, for general graphs, it belongs to the NP-hard computational complexity class. In the present paper, we propose graph reduction that enable the identification of new polynomially solvable CD sub-problem. Specifically, we show that if a graph is (*butterfly*, *diamond*)-free then a cluster deletion solution can be found in polynomial time on that graph.

1 Introduction

Clustering is a central task which can be useful for data analysis and graph mining [21]. From a graph-theoretic point of view, a cluster graph is a vertex-disjoint union of cliques [6], or equivalently, a graph which does not contain any induced P_3 subgraph, where P_3 denotes a path composed of three vertices and two edges. Clustering problems consist in making the fewest changes to the vertex and/or to the edge set of an input graph so as to obtain a cluster graph. There exist several clustering variants, including Cluster Completion (CC), Cluster Deletion (CD) and Cluster Editing (CE). In the Cluster Completion variant, edges can only be added, while, in the Cluster Deletion variant, edges can only be deleted. In the Cluster Editing, both edge additions and edge deletions are allowed. In simple terms, the goal of clustering is to partition the vertices into subsets, called clusters, which are intended to group together similar vertices. Clustering may also be useful for real-world applications issued from numerous fields including wireless sensor networks (WSN). Naturally, grouping sensor nodes into clusters has been widely adopted by the research community to satisfy the above scalability objective and generally achieve high energy efficiency and extend network lifetime in large-scale WSN environments [2]. This problem has also motivations from computational biology [3], image processing [4], VLSI design [5], and many more.

In this paper, we deal with the cluster deletion variation (CD), which may informally be defined as follows: given a simple graph, the goal is to determine the nearest cluster graph to the input graph by only deleting edges. CD

© Springer Nature Switzerland AG 2019
R. Lee (Ed.): SNPD 2018, SCI 790, pp. 76–88, 2019.
https://doi.org/10.1007/978-3-319-98367-7_7

is known to be NP-hard for general graphs [6]. However, it may become easier and even polynomial-time solvable on special graphs, for instance, when dealing with split graphs, block graphs, proper interval graphs, co-graphs [13]. CD is also polynomial-time solvable on a subclass of P_4-sparse graphs [14] that strictly includes P_4-reducible graphs, which are, in turn, a superclass of co-graphs. Those results were obtained for non weighted graphs. For weighted graphs, the cluster deletion problem can be solved in polynomial time on the class of K_3-free graphs, for which CD is equivalent to maximum weighted matching [15]. On the other hand, there are several works showing that CD is NP-hard on some subclasses of weighted graphs such as, (C_5,P_5)-free graphs (see Fig. 1), $(2K_2,3K_1)$-free graphs (see Fig. 2) and $(C_5,P_5,\text{bull},4\text{-pan},\text{fork},\text{co-gem},\text{co-4-pan})$-free graphs [13] (see Fig. 3).

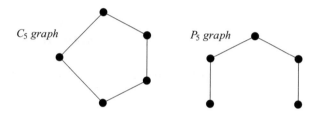

Fig. 1. Forbidden subgraphs in the class of (C_5,P_5)-free graphs

Fig. 2. Forbidden subgraphs in the class of $(2K_2,3K_1)$-free graphs

The focus of this paper is to identify new polynomial CD sub-problem based on graph reduction. In the literature, there are several graph reductions, among them the conic reduction [7], the simplicial vertices removal [9], the clique reduction [8], the C-reduction [10], the graph reduction for QoS Prediction [11], the SWR reduction [12]. All these reductions may simplify combinatorial problems on graphs. We also proposed a graph transformation, based on the *Gallai graph* [1], which enabled the identification of a new polynomially solvable sub-problem of CD on $(kite,house,xbanner,diamond)$-free graphs (see Fig. 4) [20].

In the present paper, we introduce a new reduction that relies on the neighbourhood relation between maximal cliques in (butterfly,diamond)-free graphs and establishes the tractability of CD on these graphs.

The paper is organized as follows. In the next section, we recall some notions from graph theory and define the problem we are dealing with, that is cluster

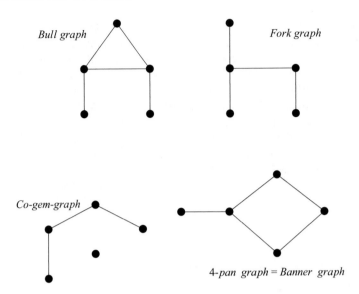

Fig. 3. Some of forbidden subgraphs in the class of $(C_5, P_5, \text{bull}, 4\text{-pan}, \text{fork}, \text{co-gem}, \text{co-}4\text{-pan})$-free graphs

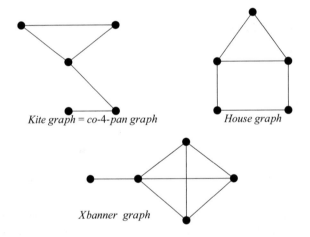

Fig. 4. Some of forbidden subgraphs in the class of (kite, house, xbanner, diamond)-free graphs

deletion. In Sect. 3, we introduce a new graph reduction based on maximal cliques and show how it can be used to prove that CD is polynomial-time solvable in (butterfly, diamond)-free graphs. In Sect. 4, we present practical algorithms, based on existing works, to recognize (butterfly, diamond)-free graphs.

2   Definitions and Notations

A graph is a mathematical structure consisting of a set of vertices and a set of edges connecting these vertices. There are several types of graphs, among them, one can distinguish simple graphs. These are graphs defined by an ordered pair (V, E), where V is a finite set of vertices and $E \subseteq \binom{V}{2}$ is a set of edges, where $\binom{V}{2}$ denotes the set of all unordered pairs of V. From a simple graph $G = (V, E)$, one can extract a spanning subgraph $G' = (V, E')$ by removing some of the edges of G, we therefore have $E' \subseteq E$. Starting from any subset U of V, one can also extract the graph $(U, E(U))$, which is the subgraph of G induced by U, where $E(U) = E \cap \binom{U}{2}$. Accordingly, a graph $G_s = (V_s, E_s)$ is a subgraph of a graph $G = (V, E)$ if there exists $U \subseteq V$ such that G_s is the subgraph of G induced by U, i.e. $V_s = U$ and $E_s = E(U)$.

A complete graph is a simple graph in which every pair of distinct vertices is connected by a unique edge. The complete graph on n vertices is denoted by K_n. The neighbourhood of a vertex $v \in V$, denoted $N(v)$, is the subset of vertices adjacent to v, i.e. $N(v) = \{u \in V \mid v\text{-}u \in E\}$, where u-v denotes the edge connecting vertices u and v. A clique of a simple graph G is a complete subgraph of G. A *maximal clique* is a clique that cannot be extended by including one more adjacent vertex. The P_3 graph is the path on three vertices as it is illustrated in Fig. 5. Observe that any complete graph cannot contain any P_3 as an induced subgraph. Henceforth, a P_3 graph composed by vertices u, v and w such that v is connected by an edge to both u and w, will be denoted by u-v-w.

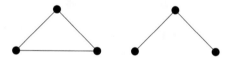

Fig. 5. The K_3 graph - The P_3 graph

In other respects, let C be a collection of (small) graphs. A simple graph G is said to be *C-free* if it contains no member of C as an induced subgraph. Many well know graph families are defined in this fashion; let us mention for instance K_3-free graphs, P_3-free graphs, claw-free graphs. To ensure that an input graph is C-free, we use recognition algorithm which generally aim to provide a reasonable answer which specify if it is C-free or not.

Formally, CD can be defined as follows

Definition 1. An *optimal solution to a CD instance defined by a graph $G = (V, E)$ is a P_3-free spanning subgraph (V, E') of G that maximizes $|E'|$.*

Therefore, an optimal solution for CD is obtained by removing, from the initial graph, as few edges as possible in order to obtain a P_3-free graph (or cluster graph). This problem belongs to the class of edge modification problems, in which one has to minimally change the edge set of a graph so as to satisfy a certain property [16,23].

3   (butterfly,diamond)-Free CD Is Polynomially Solvable

In this section, we identify a new class of graphs on which CD can be solved in polynomial time. This is the class of (butterfly,diamond)-free graphs, which contains all graphs that does not admit the two graphs shown in Figs. 6 and 7 as induced subgaphs. We propose an algorithm that efficiently solves CD for (butterfly,diamond)-free graphs based on the tractability of the maximal cliques enumeration problem on the super-class of diamond-free graphs. Indeed, a crucial property of the diamond-free class is that the maximal cliques of the graphs in this class can be enumerated is polynomial-time [17,19]. In what follows, we use this result to show that CD is tractable on graphs belonging to the (butterfly,diamond)-free class.

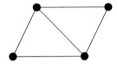

Fig. 6. Diamond graph

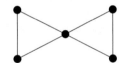

Fig. 7. Butterfly graph

As it has been shown in [22], a graph G is diamond-free if and only if it has the *one-edge-overlap* property, which means that every edge of G is contained in, at most, one maximal clique. Starting from this property, we show the following lemma

Lemma 1. *Let G be a (butterfly,diamond)-free graph, then any vertex of G appears in, at most, one maximal clique of G whose size is three or more.*

Proof. Let C and C' be two maximal cliques in G having cardinality three or more. Suppose that there exists a vertex, or more, which appear in both C and C', and proceed to get a contradiction. We distinguish the two following cases:

- $|C \cap C'| = 1$: this means that the maximal cliques C and C' share exactly one vertex, say v. Since C and C' have a size three or more, each should contains at least, two other vertices. Assume, therefore, that $\{t, u, v\} \subseteq C$ and $\{v, w, x\} \subseteq C'$. Since C and C' are maximal cliques, there must exist $y \in C$ such, that $\{x, y\} \notin E$. It follows that $(\{t, y, v, w, x\}, E(\{t, y, v, w, x\}))$ has an induced butterfly or an induced diamond, which contradicts the hypothesis.

– $|C \cap C'| \geq 2$: this means that C and C' share, at least, two common vertices, say v and w. Moreover, we have $\{v, w\} \in E$, which means that C and C' share a common edge. This contradicts the *one-edge-overlap* property for diamond-free graphs (see Proposition 1 of [22]).

Solving CD on a (butterfly,diamond)-free graph can be split into two phases. The first phase consists in calculating all maximal cliques of G that have size three or more. By Lemma 1, the number of such maximal cliques in a (butterfly,diamond)-free graph with vertex set V cannot exceed $|V|$. Moreover, we show that all such cliques must be part of every CD solution.

Lemma 2. *Let G be a (butterfly,diamond)-free graph. Then any CD solution for G contains all cliques of size three or more.*

Proof. Let $G^* = (V, E^*)$ be a CD solution for G. Assume that there exists a maximal clique, say C, such that $|C| \geq 3$ and $E(C) \nsubseteq E^*$, where $E(C)$ denotes the edges of E whose both endpoints are in C, then proceed to get a contradiction.

If $E(C)$ is not entirely included in E^*, then C may be uniquely bi-partitioned into two subsets C_{in} and C_{out} such that C_{in} is the smallest subset of C that verifies the followings

$$E(C_{in}) = E(C) \cap E^* \qquad \text{and} \qquad E(C_{out}) \cap E^* = \varnothing \qquad (1)$$

This choice of C_{in} and C_{out} implies the followings facts:

(*i*) C_{out} cannot be empty, otherwise, $E(C)$ will be entirely included in E^*.
(*ii*) If C_{in} is not empty then it should contain, at least, two vertices.
(*iii*) Any edge with one endpoint in C_{in} and the other one in C_{out} is not in E^*.
(*iv*) Let $p = |C_{in}|$ and $q = |C_{out}|$. We have therefore $p + q \geq 3$ and $q \geq 1$.

Next, we prove that E^* cannot contain an edge $\{u, v\}$ having exactly one endpoint, say u, which is precisely in C_{in}. Suppose the converse it true. First, observe that, by (*iii*), v cannot be in C_{out}. And since it is not in C_{in}, we deduce that $v \notin C$. By (*ii*), this will imply that $E(C_{in})$ contains, at least, one edge, $\{u, u'\}$. Recall that u is already in a maximal clique having size three or more, that is C. On the other hand, $\{u, v\}$ is in E^*. Furthermore, $\{u, u'\}$ is also in E^* since $\{u, u'\} \in E(C_{in}) \subseteq E^*$. But, (V, E^*) is P_3-free since it is a CD solution. This implies that $\{u', v\} \in E^*$. We deduce that $\{u, u', v\}$ is a clique. It must, therefore, be comprised in a maximal clique, say C'. Moreover, we have $C' \neq C$ since v is in C' but not in C. As a result, u belongs to two distinct maximal cliques having size three or more, that are C and C'. This contradicts Lemma 1.

The following step is to prove that every vertex of C_{out} cannot be the endpoint of more that one edge of E^*. Suppose the converse is true, which implies that there exists $u \in C_{out}$ such that $\{u, v\}, \{u, v'\} \in E^*$, where $v \neq v'$. First, observe that, by the choice of C_{in} and C_{out}, v and v' cannot be in C. Moreover, since (V, E^*) is P_3-free, we deduce that $\{v, v'\} \in E^*$. Thus, $\{u, v, v'\}$ is a clique, and then it must be included in a maximal clique that differ from C, because $v, v' \notin C$. But this latter clique, which has size three or more, share vertex u with clique C. This contradicts Lemma 1.

As a consequence, the number of edges in E^* that have exactly one endpoint in C cannot exceed $|C_{out}| = q$.

Consider, therefore, the spanning subgraph H obtained from G^* by including all the edges of $E(C)$ and eventually withdrawing the edges of E^* that have exactly one endpoint in C. To compare the cost of H against that of G^*, we take into account the following observations:

- The number of edges that can be withdrawn is bounded by q.
- The number of edges in $E(C)$ that have one endpoint in C_{in} and the other one in C_{out} is equal to pq.
- The number of edges in $E(C_{out})$ is $q(q-1)/2$.

The cost of H can be bounded from above as follows:

$$\text{cost}(H) \geq \text{cost}(G^*) - q + pq + \frac{q(q-1)}{2}$$

Since G^* is an optimal solution, we must have $\text{cost}(H) \leq \text{cost}(G^*)$. This implies that:

$$q \geq pq + \frac{q(q-1)}{2} \tag{2}$$

Inequality (2) and $q \geq 1$ imply that $p = 0$ or $p = 1$. If $p = 0$ then, using $p + q \geq 3$, we obtain $q \geq 3$, which implies that $q < q(q-1)/2$ and contradicts (2). Otherwise, ($p=1$), using $p + q \geq 3$, we obtain $q \geq 2$, which implies that $q(q-1)/2 \geq 1$ and contradicts (2). We conclude that every optimal solution must contain all cliques having size three or more.

The second phase of the CD solution algorithm dedicated to (butterfly, diamond)-free graphs consists in extending the partial solution made up of all cliques of size three or more, by as much two-cliques as possible. This can be done by resorting to the calculation of a maximum matching in the subgraph induced by $V \setminus \bigcup_{i=1}^{m} C_i$, where the C_i's are the maximal cliques of G that have size three or more.

Example 1. *Figures 9 and 10 illustrates how the two-phase algorithm operates on the graph of Fig. 8. It is easy to see that this graph is (butterfly, diamond)-free. In the first phase, two maximal cliques of sizes three and another of size four are detected.*

In the second phase, starting from the residual graph, a maximum matching is calculated. The CD solution is therefore obtained by putting together the cliques obtained in the first phase and the maximum matching (see Fig. 11).

Theorem 1. *Let G be a (butterfly, diamond)-free graph. Then an optimal CD solution for G can be found in polynomial time.*

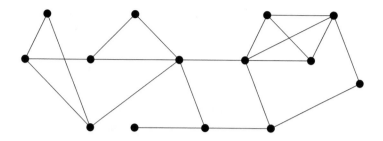

Fig. 8. A graph.

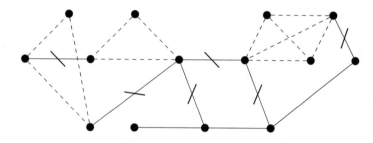

Fig. 9. Application of two-phase algorithm: phase 1

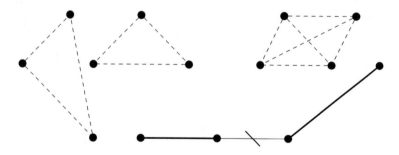

Fig. 10. Application of two-phase algorithm: phase 2

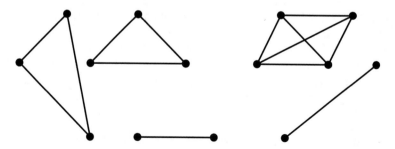

Fig. 11. An optimal CD solution for the (butterfly,diamond)-free graph of Fig. 8 obtained by means of the two-phase algorithm.

Proof. Let $C_1, \ldots, C_m$ be the maximal cliques of $G = (V, E)$ that have size three or more. Then define the vertex subset $\hat{C}$ and the edge subset $\hat{E}$ as follows:

$$\hat{C} = \bigcup_{i=1}^{m} C_i \quad \text{and} \quad \hat{E} = \bigcup_{i=1}^{m} E(C_i)$$

Let M be a maximum matching of the subgraph of G induced by $V \backslash \hat{C}$. Define, therefore, the graph G^* as $(V, \hat{E} \cup M)$. We show that G^* is an optimal CD solution for G.

Clearly, G^* is a spanning subgraph of G, because $\hat{E} \subseteq E$ and M is a matching of a subgraph of G, which implies that $M \subseteq E$. Next, we prove that G^* is P_3-free. Suppose that G^* has a P_3, u-v-w, as an induced subgraph and proceed to get a contradiction.

The hypothetical induced P_3 requires that both $\{u, v\}$ and $\{v, w\}$ are in $\hat{E} \cup M$, but not $\{u, w\}$. By Lemmas 1 and 2, $\hat{E}$ is the union of edges forming vertex-disjoint cliques, (the C_i's). Then $\{u, v\}$ and $\{v, w\}$ cannot be simultaneously in $\hat{E}$, since $\{u, w\} \notin \hat{E}$. Similarly, M is a matching, that is, a set of non incident edges. Then $\{u, v\}$ and $\{v, w\}$ cannot be simultaneously in M. It remains to check whether it is possible to have $\{u, v\} \in \hat{E}$ and $\{v, w\} \in M$, (or the reverse). This implies that v is in $\hat{C}$ and v is in $V \backslash \hat{C}$, which is not possible.

The last step consists in proving that G^* is an optimal solution. Suppose there exist a spanning P_3-free subgraph, $G' = (V, E')$, of G such that $|E'| > |\hat{E} \cup M|$.

By Lemma 2, G' must contain all cliques of G having size three or more, i.e., the C_i's. It follows that E' contains all the edges involved in these cliques, i.e., the elements of $\hat{E}$. Then, we can write $E' = \hat{E} \cup M'$. Moreover, the elements of M' are edges of the subgraph of G induced by $V \backslash \hat{C}$. Indeed, the endpoints of every edge e' of M' must be vertices of $V \backslash \hat{C}$, because, otherwise, e' will intersect with a C_i, and this contradicts the fact that G' is a P_3-free graph. Moreover, the edges of M' must be pairwise non incident edges, because these edges cannot form any clique of size three or more and G' is a P_3-free graph. It follows that M' is a matching of the subgraph of G induced by $V \backslash \hat{C}$. This implies that $|M'| \leq |M|$, since M is a maximum matching of the same subgraph, that is, the subgraph of G induced by $V \backslash \hat{C}$. This results in a contradiction with the hypothesis.

As a last step of the proof, we show that G^* can be constructed in polynomial time. It is well established that the problem of enumerating all maximal cliques in a simple graph can be achieved in $O(|V|^2 \kappa)$, where κ denotes the number of maximal cliques in G. By the *one-edge-overlap* property, κ cannot exceed $|E|$ in (diamond)-free graphs. It follows that the cliques of G can be extracted in $O(|V|^2 |E|)$. In addition, there is a $O(|V|^2 |E|)$ time algorithm to find a maximum matching in a graph [15]. It follows that CD is polynomial-time solvable in the class of (butterfly,diamond)-free graphs.

The first phase comprised in the two-phase algorithm requires finding all maximal cliques. To achieve this task, we use the *BronKerbosch algorithm* [17], which is one of the most successful maximal clique enumeration algorithms. It is a simple backtracking procedure that recursively enumerates all cliques whose size is bounded by a parameter of the algorithm [17,18]. This algorithm has already been presented in many versions. So, we briefly outline the classic version of this algorithm [24,25] (see Algorithm 1).

Algorithm 1. Classic BronKerbosch algorithm

---

1 **Input:** R, P, X
2 **Output:** List of maximal cliques of G
3 **Begin**
4 | **If** P *and* X *are both empty* **then**
5 | | report R as a maximal clique
6 | **end if**
7 | **For** *each* $v \in P$ **do**
8 | | BronKerbosch($R \cup v$, $P \cap N(v)$, $X \cap N(v)$)
9 | | $P \longleftarrow P \setminus \{v\}$
10 | | $X \longleftarrow X \cup \{v\}$;
11 | **end for**
12 **End**

On the first call R and X are both empty, and P contains all the vertices of the graph. Let R be the temporary result, P be the set of the possible candidates and X be the excluded set. $N(v)$ indicates the neighbors of the vertex v. The algorithm pick a vertex v from P to expand. After that, add v to R and remove its non-neighbors from P and X. Then pick another vertex from the new P set and repeat the process. Continue until P is empty. Once P is empty, if X is empty then report the content of R as a new maximal clique. If it's not then R contains a subset of an already found clique. Now, backtrack to the last vertex picked and restore P, R and X as they were before the choice, remove the vertex from P and add it to X, then expand the next vertex. If there are no more vertices in P then backtrack to the superior level.

The complexity of the BronKerbosch algorithm is polynomial in the number of maximal cliques which is in turn bounded by the number of edges in (butterfly,diamond)-free graphs.

4   Recognition of (butterfly,diamond)-Free Graphs

The goal of this section is to recognize (butterfly,diamond)-free graphs. To this end, we present two algorithms tailored for the recognition of the graphs belonging in this class. Whenever an input graph is recognized as (*diamond, butterfly*)-free, a cluster graph can be extracted from this graph using the two-phase algorithm.

Recognizing (butterfly,diamond)-free graphs can be done by focusing, in turn, on the neighbourhood of each vertex of the input graph. We proceed by checking the absence of the forbidden graphs, i.e. the butterfly and the diamond, separately.

We start from the observation that a graph is diamond-free if and only if the neighbourhood of every one of its vertices induces a P_3-free graph (Fig. 5). Hence, the role of Algorithm 2 is to check if this condition is satisfied for every vertex of the input graph.

Algorithm 2. Algorithm to recognize diamond-free graphs

---

1 **Input:** graph $G = (V, E)$
2 **Output:** G is diamond-free or not
3 **Begin**
4   **For all** $v \in V$ **do**
5        $G_v = (N(v), E(N(v))$ /* graph induced by $N(v)$.*/
6   **For all edges** $\{u, w\}, \{w, x\} \in E(N(v))$ **do**
7   **If** $A_G[u, x] = 0$ **then**
8   return FALSE
9   **end if**
10   **end for**
11   **end for**
12   return TRUE
13 **End**

In turn, Algorithm 3 checks the input graph against butterfly induced subgraphs. We use the fact that a graph is butterfly-free if and only if the subgraphs induced by the neighbourhood of every vertex is $2K_2$-free (Fig. 2). Both recognition algorithms run in $O(|V||E|^2)$ and use an adjacency matrix, denoted by A_G, to implement an input graph G.

Algorithm 3. Algorithm to recognize butterfly-free graphs

---

1 **Input:** graph $G = (V, E)$
2 **Output:** G is butterfly-free or not
3 **Begin**
4   **For all** $v \in V$ **do**
5        $G_{N(v)} = (N(v), E(N(v))$ /*graph induced by $N(v)$ */
6   **For all non incident edges** $\{u, w\}, \{x, y\} \in E(N(v))$ **do**
7   **If** $A_G[u, x] = 0$ *and* $A_G[u, y] = 0$ *and* $A_G[w, x] = 0$ *and* $A_G[w, y] = 0$ **then**
8   return FALSE
9   **end if**
10   **end for**
11   **end for**
12   return TRUE
13 **End**

---

5   Conclusion

In this paper, we identified a polynomial-time solvable class of the cluster deletion problem (CD). We presented a two-phase algorithm, which relies on the adjacency between maximal cliques in (butterfly,diamond)-free graphs. The first phase of the proposed algorithm consists in calculating all maximal cliques having a size of three or more. This results in a partial solution, which is, subsequently, extended by the edges of a maximum matching of a sparse spanning subgraph. The proposed two-phase algorithm and its correctness proof show that CD is polynomial-time solvable in (butterfly,diamond)-free graphs.

Besides, we presented practical algorithms, based on existing works, to recognize (butterfly,diamond)-free graphs that characterize the identified polynomial-time solvable classes. We showed that we can recognize butterfly-free and diamond-free graphs in $O(|V||E|^2)$. These results show CD is tractable on (butterfly,diamond)-free graphs.

Work is in progress in order to derive wider classes of cluster deletion problem by targeting graphs emerging in specific applications and forbidding induced subgraphs that are rarely induced in the targeted graphs.

References

1. Le, V.B.: Gallai graphs and anti-Gallai graphs. Discrete Math. **159**(1–3), 179–189 (1996)
2. Mamalis, B., Gavalas, D., Konstantopoulos, C., Pantziou, G.: Chapter 12: Clustering in Wireless Sensor Networks, RFID and Sensor Networks: Architectures, Protocols, Security, and Integrations (2009)
3. Zahn, C.T.: Approximating symmetric relations by equivalence relations. J. Soc. Ind. Appl. Math. **12**, 840–847 (1964)
4. Pipenbacher, P., Schliep, A., Schneckener, S., Schnhuth, A., Schomburg, D., Schrader, R.: ProClust: improved clustering of protein sequences with an extended graph-based approach. Bioinformatics **18**(Suppl. 2), 182–191 (1964)
5. Guo, J.: A more effective linear kernelization for cluster editing. Theoret. Comput. Sci. **410**, 718–726 (2009)
6. Shamir, R., Sharan, R., Tsur, D.: Cluster graph modification problems. Discrete Appl. Math. **144**, 173–182 (2002)
7. Vadim, V.L.: Conic reduction of graphs for the stable set problem. Discrete Math. **222**(1–3), 199–211 (2000)
8. Lovász, L., Plummer, M.D.: Matching Theory. Akadémiai Kiadó, vol. 121 of the North-Holland Mathematics Studies, North-Holland Publishing, Amsterdam (1986)
9. Brandstädt, A.L., Hammer, P.: On the stability number of claw-free P5-free and more general graphs. Discrete Appl. Math. **95**(1–3), 163–167 (1999)
10. Catlin, P.A.: A reduction method for graphs. Congressus Numerantium **65**, 159–170 (1988)
11. Goldman, A., Ngoko, Y.: On graph reduction for QoS prediction of very large web service compositions. In: International Conference on Service Computing, Honolulu, US (2012)

12. Cardoso, J., Sheth, A., Miller, J., Arnold, J., Kochut, K.: Quality of service for workflows and web service processes. Web Semant. Sci. Serv. Agents World Wide Web **1**(3), 281–308 (2004)
13. Gao, Y., Hare, D.R., Nastos, J.: The cluster deletion problem for cographs. Discrete Math. **313**(23), 2763–2771 (2013)
14. Bonomo, F., Durn, G., Napoli, A., Valencia-Pabon, M.: A one-to-one correspondence between potential solutions of the cluster deletion problem and the minimum sum coloring problem, and its application to P_4-sparse graphs. Inform. Process. Lett. **115**, 600–603 (2015)
15. Edmonds, J.: Maximum matching and a polyhedron with 0,1-vertices. J. Res. Natl. Bur. Stand. B Math. Math. Phys. **69B**, 125–130 (1965)
16. Natanzon, A., Shamir, R., Sharan, R.: Complexity classification of some edge modification problems. Discrete Appl. Math. **113**, 109–128 (2001)
17. Bron, C., Kerbosch, J.: Finding all cliques of an undirected graph (Algorithm 457). Commun. ACM **16**(9), 575–576 (1973)
18. Cazals, F., Karande, C.: A note on the problem of reporting maximal cliques. Theor. Comput. Sci. **407**(1–3), 564–568 (2008). https://doi.org/10.1016/j.tcs.2008.05.010
19. Tomita, E., Kameda, T.: An efficient branch-and-bound algorithm for finding a maximum clique with computational experiments. J. Global Optim. **37**(1), 95–111 (2007)
20. Malek, S., Naanaa, W.: A new polynomial cluster deletion subproblem. In: 13th IEEE/ACS International Conference of Computer Systems and Applications, pp. 1–8 (2016)
21. Ur Rehman, S., Ullah, K.A., Fong, S.: Graph mining: a survey of graph mining techniques. In: International Conference on Digital Information Management, ICDIM, pp. 88–92 (2012). https://doi.org/10.1109/ICDIM.2012.6360146
22. Fellows, M.R., Guo, J., Komusiewicz, C., Niedermeier, R., Uhlmann, J.: Graph-based data clustering with overlaps. Discrete Optim. **8**, 2–17 (2011)
23. Bossard, A.: The decycling problem in hierarchical cubic networks. J. Supercomputing **69**(1), 293–305 (2014)
24. David. E., Maarten, L., Darren, S.: Listing all maximal cliques in sparse graphs in near-optimal time. In: Exact Complexity of NP-hard Problems, 31 October–05 November 2010 (2010)
25. Alessio C., Review of the Bron-Kerbosch algorithm and variations, report, School of Computing Science, Sir Alwyn Williams Building, University of Glasgow (2013)

A Study on the Lie Detection of Telephone Voices Using Support Vector Machine

Hyungwoo Park[1(✉)], Jong-Bae Kim[2], Seong-Geon Bae[3], and Myung-Sook Kim[4]

[1] School of Information Technology, Soongsil university, Seoul, Republic of Korea
pphw@ssu.ac.kr
[2] Department of Telecommunication Engineering, Soongsil University, Seoul, Republic of Korea
kjb123@ssu.ac.kr
[3] Sori Engineering Lab, Computer Media Information Engineering, Kangnam University, Gugal-dong, Giheung-gu, Yongin-si, Gyeonggi-do, Korea
sgbae@kangnam.ac.kr
[4] Department of English Language and Literature, Soongsil University, Seoul, Republic of Korea
kimm@ssu.ac.kr

Abstract. Human voices are one of the easiest ways to communicate information between humans. Voice characteristics may vary from person to person and include voice rate, genital form and function, pitch tone, language habits, and gender. Human voices are a key element of human communication. In the era of the Fourth Industrial Revolution, the voices are the main means of communication between people and people, between humans and machines, machines and machines. And for that reason, people are trying to communicate their intent clearly to others. In the process, language information and various additional information are included. Information such as emotional state, health status, reliability, presence of lies, changes due to alcohol, etc. These languages and non-linguistic information can be used as a device to assess the lie of telephone voices that appear as various parameters. Especially, it can be obtained by analyzing the relationship between the characteristics of the fundamental frequency (fundamental tone) of the vocal cords and the resonance frequency characteristics of vocal tracks. Previous studies have extracted parameters for false testimony of various telephone voices and conducted this study to evaluate whether a telephone voice is a lie. In this study, we proposed a judge to judge whether a lie is true by using a support vector machine. We propose a personal telephone truth discriminator.

Keywords: Voice analysis · Lie detection · Credit evaluation Telephone voice

© Springer Nature Switzerland AG 2019
R. Lee (Ed.): SNPD 2018, SCI 790, pp. 89–99, 2019.
https://doi.org/10.1007/978-3-319-98367-7_8

1 Introduction

A voice of the person is one of the convenient and easiest ways to exchange information people together and men and machines. These voices are made in the human vocal tract and spread through the air. The voices made by vocal institutions are very young, so they have learned so much that they can communicate easily and habitually. And the voices that are created to spread far and disappear. At this time, the voices produced are made with the information necessary for verbal communication, the voice is created by the personality of the person who generate it [1]. These individual characteristics are similar but different parameters are obtained by analyzing the differences between the characteristics of the vocal organs, the spatial characteristics of the personality, the language, the psychological situation and the health state. In the voice analysis, similar features of sound character are meaning of language, and different features are caused by individual habits. As people's voices have been easily used in various aspects of information transmission, in information technologies, have been developed to record or to be transmitted far away. Today, this voice analyzes the information of the voices, They record information, judge psychology, and set the direction of people and machines [1–3].

In the process of being created, speech has parameters that give different characteristics to each individual. This is a different sound, because the shape and characteristics of vocal organs are different, that depends on the sound generation process and resonance characteristics of personality. In addition, the vocabulary used, speech habit and the characteristics of the location, are differentiated according to the difference of the residential area. Voice characteristics also change according to health and psychological condition. So, by analyzing these voices in detail, we can discover the differences. Thus, the generated sound can be analyzed by a person or a machine, and the result basically conveys the meaning of the language. In addition to basic language information, you can analyze and identify more than 100 additional information to identify health and psychological conditions, and can be used to determine the reliability or deceit of a speaker [3, 4].

With the development of information devices, you can process and analyze your voice and get various results. First, there is a method to transfer voice from a communication device or to save voice using a storage device. Next, the meaning of the voice is analyzed, and instructions and operations of the machine can be performed through speech recognition. Next, people can recognize people by telling them in some form (speaker recognition), allowing access through identification (speaker identification), or by determining whether important functions are performed. The computer can also aggregate voice prompts or send information from the device to a person [5, 6]. Today, when using technologies such as communication, storage, speech recognition, speaker identification, speaker recognition, and speech synthesis, it is difficult to handle the speaker's feelings together. However, as in [5, 6], a study has been attempted to add the properties of synthetic language and add emotional parts of speech synthesis as in [6]. In [5], attempts have been made to distinguish the amount and type of information that recognizes and communicates emotions in speech recognition and communication [7].

In the 4th Industrial Revolution ear, artificial intelligence technology is developing and affecting society as a whole. In particular, deep learning is a popular kind of method of machine learning that is often performed by a self-study. And, adding features, around the Google Deepmind challenge match on March 2016 [6]. And the machine learning method combined with the technique of processing big data, has been able to deal with areas that were previously impossible with artificial intelligence. A big data processing method known as deep turging is a machine learning algorithm that can combine various nonlinear transformation techniques into a computer to perform multi-step abstraction (key parameter combination of big data) [1]. However, the machine learning method requires a large amount of data, a computation ability to make a conclusion, and a resource to judge the result [1, 3, 4]. In this study, we used the support vector machine, which can obtain effective results even in a small amount of learning, to distinguish the characteristics of voices.

SVM (Support Vector Machine) is one of the machine learning fields and is a supervised learning method for pattern recognition and data analysis [8]. Machine learning can be divided into supervised learning and non-supervised learning [9]. The advantage of supervised learning is that even small calculations can make quick decisions because the predictions are relatively accurate and use a small amount of data [10]. SVM is a method of classifying data by finding a linear decision boundary (hyperplane) that distinguishes all data elements of one class from data elements of another class [11]. And this SVM has very high classification accuracy [8]. The reason for the high accuracy is that the margins between the data points to be classified are maximized and the problem of overfitting occurs less frequently [9]. In addition, if it is difficult to determine by a linear classification algorithm, it is easy to use kernel functions to improve judgment performance [8, 9].

People are trying to make a good voice for a variety of reasons. The purpose is to clearly communicate the intentions to those who are preparing for various speeches, personal conversations, sales, interviews about their occupations and clearly communicate the intended situation. And people use financial institutions for a variety of reasons. The basic function of finance is to enable households, corporations, governments, financial institutions, etc. to obtain necessary funds and manage funds through transactions. Credit of individuals and institutions of financial institutions is very important information. The criteria for evaluating creditworthiness are the extent to which past performance and format are faithfully implemented, the willingness to repay, if any, and the probability of default. In the field of finance and P2P finance, it is very popular because it adopts the latest technologies such as in-depth learning and psychological evaluation based on credit rating. In this study, we propose an algorithm of lie detection of personal telephone speech, that is quick, easy to use and accurate through sound analysis. In previous studies, we studied the parameters of the voice as a function of credit change. Previous studies have also examined changes in the voice parameters before and after bankruptcy in relation to the default of loans, and studies have been conducted to further characterize these changes. We also propose a personal credit evaluation module by analyzing the telephone voice and using these parameters and Machine Learning (SVM) to determine the falsehood of the telephone voice.

In Sect. 2, we have looked the basic voice generation and analysis, and support vector machines. In Sect. 3, we introduce the propose method for evaluating the telephone voice. In Sect. 4, the experiment and the result. And in Sect. 5, we conclude.

2 Related Works and Basic Algorithm Reviews

2.1 Speech Analysis

The speech communication is a technology of information transmission that has been used for a long time. The process by which the speaker tries to convey the meaning and the celestial understanding is basically started from the concept that the speaker intends to deliver, and the process is as follows. The speaker changes that idea through the language structure and selects the appropriate word or word to represent the speaker's thoughts in the process. And then arranges the order of words according to the grammar, and performs processing that emphasizes highlighting, emphasis, tonal changes, and pitch, formant changes due to habits or dialects. And the next step, their brain command is issued that moves the position of the vocal organs and the muscle tissue associated from the vocalists desire pronunciation. This command is prepared in the voice organ and the airflow from the lungs vibrate the vocal cords. With this vibration and airflow resonates with the vocal track spreading to the nose and mouth. Then, the acoustic waveforms corresponding to the intention of the speaker are generated [10, 11]. The Fig. 1 shows process of speech communication.

In speech acoustic signal processing, the information of the voice can be largely classified into the characteristics of the excitation source and the characteristics of the vocal track parameter. First, the characteristics of the excitation source can be confirmed by the presence or absence of vibration of the vocal cords, and the fundamental frequency that occurs when the vocal cords vibrate is called the pitch. This can be determined by analyzing the number of vibrations during a unit time or during the period during which the vocal cord is opened and closed. When the pitch is accurately detected, the influence of the speaker at the time of speech recognition can be reduced, and the probability of speech synthesis can be maintained or the naturalness can be easily maintained. And if we know exactly this pitch, we can use it to change to another voice. For a typical male, the available pitch range is 80 to 250 Hz. In women, there are characteristics between 150 and 300 Hz [10, 12]. The change in pitch over time can be regarded as a parameter of major change in speech, and the change in pitch in addition to the language information included in the voice can be regarded as a method of estimating other information. Second, the vocal track parameter is formant. The formant frequency of the voice is the frequency band emphasized by the resonance when the air tremor that occurs in the vocal cords passes through the vocal track. This formant frequency is represented by the first and second formants, or F1 and F2, in order from lowest frequency. The formants generally have first to fourth degree of resonance, and the fifth and sixth formants are also detected when the vocal tract is good. The position of the vertex is indicated by the frequency value of the vertex. And if you look at the formant's bandwidth, you can see what kind of vocal track it has.

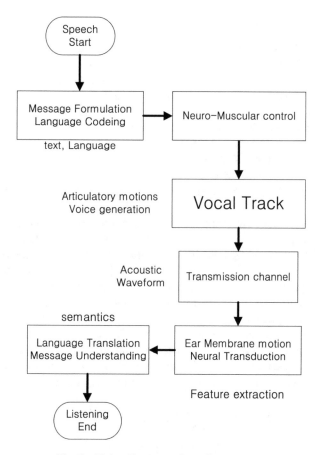

Fig. 1. Voice Genetraon flow diagram [11]

Although the sound due to the air quenching occurring in the excitation is the same, the emphasized frequency band varies with thickness, length and rate of change [4, 10, 12].

Generally speaking, the phonetic characteristics according to the phoneme are represented by F1 and F2, that is the first and the second formant. And F3, F4, and F5 represent the individual characteristics of the speaker. At this time, frequency position, bandwidth, amplitude, etc. of F3 and F4 can be classified into characteristics. When voice recognition is performed, F1 or F2 is important information in the voiced section. However, in the unvoiced part, since the formant part is not simple and complicated unlike the voiced part, not only F1 and F2 but also F3, F4 and F5 include phonetic information and other information. It can also be confirmed by analyzing the slope of the formant to evaluate the pronunciations and deliver the information to the listener [11, 12]. Thus, the positions and slopes of the first and second formants are obtained, and the slopes of the first and fourth formants are compared to confirm the reliability and clarity of the ignition as a whole. We can also use this formant frequency change as a parameter to measure speech [4].

2.2   Support Vector Machine

Support Vector Machine (SVM) is one of the algorithms used in machine learning analysis in the IT industry. SVM was developed based on statistical analysis, and the results of the algorithm are called dependent variables, and the factors affecting the results are called independent variables [8, 9]. Statistical analysis is a method of predicting dependent variables by predicting the statistical similarity of independent variables in a large number of data [9, 13]. In other words, it is a method to analyze various conditions and make a statistical judgment criterion based on a large number and a small number [14]. SVM is a simple algorithm for finding hyper-planes that discriminate or predict dependent variables (decision results) through feature extraction from many data [15]. "SVM is a supervised learning model with associated learning algorithms that analyze the data used for classification and regression analysis [15]. SVM is an algorithm that can distinguish two categories "as far as possible" when there are more than one category in the dependent variable [16, 17]. For example, algorithms that distinguish between ripe apples and inexperienced apples are done through machine learning and large data analysis. This SVM is well used in the field of machine learning because of its very high classification accuracy. The reason for the high classification accuracy is that it isolates the dependent variable and maximizes the error margin of the hyperplane margin [18, 19]. Usually other predictors are learning how to reduce the error, in this case overfitting and the SVM is not overly suitable [20]. It can improve prediction performance by changing the dimension of data called kernel function, and it is easy to use when the characteristics of data are well known [21]. In this study, the hyperplane was constructed by analyzing the parameters of the voice change, the bankruptcy and the voice change characteristics before and after bankruptcy. So SVM can judge. In addition, we are actively studying and changing hyperplanes during data analysis. The parameters obtained from the speech analysis mentioned in Sect. 2.1 are used as independent variables through judgment. The parameters used in the determination are as follows: a change parameter of the fundamental frequency and the pitch frequency, a slope parameter according to the formant position and size, a parameter of the speech speed according to the time change of the formant, and a slope parameter according to the energy change.

2.3   Finance and ICT Technology

A peer-to-peer loan (also known as a P2P loan or social loan) is a method of raising an individual's debt. And it is personally provided to the company through an online service that does not have a formal arbitration institution. Direct matching of lenders and lenders by removing intermediaries from the process. However, ease of use and cost savings are more risky than traditional financial institutions. P2P lenders can reduce operating costs by providing services only online. Therefore, they can provide services at a lower cost than traditional financial institutions. As a result, lenders can benefit more from financial instruments offered by traditional financial institutions, And the borrower is less likely than the existing bank. They can get loans at interest rates. The P2P platform itself can also earn money by imposing fees on successful transactions. In other words, P2P lenders do not lend their own funds to the borrower. They act as facilitators for both borrowers and investors [2, 3].

3   Telephone Voice Discrimination of Propose Methods

The proposed personal lie detection methods in telephone voice is as follows. When a voice is input by telephone conversation, the classifier prepares the voice by noise processing and then disseminate the voiced and unvoiced section of speech. The voice then extracts the parameters that SVM will use as an independent variable. Also, make sure that has other arguments, that discriminator can use and classify separately. Then enter the parameters in the SVM discriminator to perform an untruth rating over the phone voice. The result shows the authenticity of the phone call. A block diagram of the proposed technique is shown in Fig. 2. In the signal processing, a spectral sub-traction method is generally used for information that is a preprocessor for make clean speech of voice. In the feature extraction, the feature information obtained from the previous study was used that is using cepstal method. In previous studies, we identified changed speech components before and after individual bankruptcy. In this study, we use the parameter of formant slope, pitch, frequency and speech speed for deciding the individual reliability.

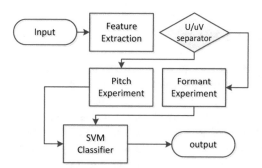

Fig. 2. Block diagram of proposed method

In the case of the pitch used as the judgment parameter of the SVM, the parameter is extracted using the mean and variance of the pitch frequency over time rather than merely analyzing the frequency. Figure 3 shows a pitch contour diagram for 'ne' (mean is yes) utterance in speaker-independent. In the figure, the abscissa is the frame order and the ordinate is the pitch frequency value. Figure 3(a) shows the voice of a normal debtor who has not defaulted. Figure 3(b) shows the results of the person who caused the default. In Fig. 3(a), the pitch change rate is low and the pitch is evenly analyzed. In Fig. 3(b), the pitch change is relatively high and the variance is expected to be high. Using these results, it is possible to judge the truthfulness of telephone voice related to finance through pitch analysis. And it can be defined as pitch dispersion as the first discrimination standard of SVM.

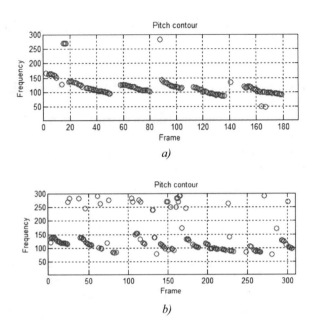

Fig. 3. Pitch contour of speeches

4 Experiment and Results

In order to obtain parameter for this study, we were allowed to use the telephone voice record files for research purposes, which is a loan borrower and counselor from Korean savings banks (S-Capital and m-bank). Then, the recorded voices were analyzed by proposed method that is parameterization, and calculation of the results. Furthermore the analysis were compared with the actual personal credit score to find the similarities between these data. The 30 speechs from the telephone conversation, comprising 18 males and 12 females, were used for data. The age group for the sample data is all in the 20s–40s. Then, the collected voices were sampled with 11 kHz and quantized to 16 bits per sample. Figures 4 and 5 shows the data analysis results as a result of the speaker-dependent analysis. Figures 4 and 5 show cepstrum analysis results for a specific section of speech signal. When cepstrum analysis is performed, the vocaltrack parameter can be obtained at low queriency and information corresponding to pitch at high queriency can be obtained. Personal credit status changes are voice data before and after the bankruptcy of the telephone consultation voice. In Fig. 4, a smooth harmonic shape appears at medium latency. However, Fig. 5 differs from Fig. 4 in shape and structure, and also differs from Fig. 3 in the shape of the distribution of pitch. You can also use some of these parameters to create the SVM Inspector baseline. And at low queriency we get information about formants and formants. And we can extract information about speech rate using a combination of queriency according to the change of time.

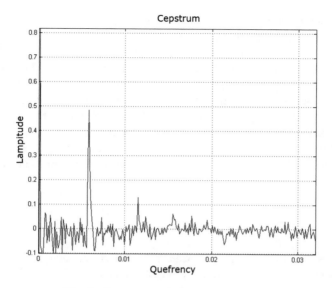

Fig. 4. Cepstrum analysis of normal voice

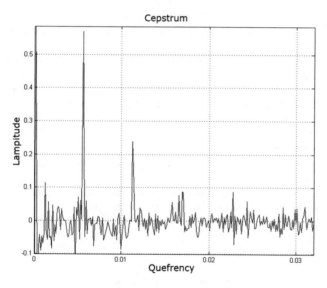

Fig. 5. Cepstrum analysis of default voice

5   Conclusion

Speech is an important method of information transfer. In addition to linguistic information, the voice contains a variety of data, such as health, emotional and honesty. The speech signal is a time-varying signal that changes various parameters over time. However the voice is quasi-periodic data, that can be able to make parameter or criteria for various other information. The changes will vary by language, but are typically three to four times per second. That is, the vocal track parameters and excitation source are changing. In the meantime, IT development has put a lot of effort into understanding the language information contained in people's words. In this study, we introduced a classifier which can evaluate voice by analyzing voice other than language information. This discriminant algorithm can be used to predict the default or normal state of speech analyzed by binary decision logic. We investigated the characteristics of each individual voice and identified the parameters of a reliable parameter. In short, a reliable voice has a broader range of tonal tones, clear pronunciation and, a clearer formant structure. We can use this parameter to create a determinant that evaluates trust.

References

1. Park, H.W.: A study on personal credit evaluation system through voice analysis: by machine learning method. In: KSII, the 9th International Conference on Internet (ICONI), Vientiane Laos, December 2017
2. Lee, S.: Evaluation of mobile application in user's perspective: case of P2P lending apps in FinTech industry. KSII Trans. Internet Inf. Syst. **11**(2), 1105–1117 (2017)
3. Park, J.W., Park, H.W., Lee, S.M.: An analysis on reliability parameter extraction using formant of voice over telephone. Asia Pac. J. Multimedia Serv. Converg. Art Hum. Sociol. **7**(3), 183–190 (2015)
4. Park, H.W., Bae, M.J.: Analysis of confidence and control through voice of Kim Jung-un's. Information **19**(5), 1469–1474 (2016)
5. Han, K., Yu, D., Tashev, I.: Speech emotion recognition using deep neural network and extreme learning machine. In: Interspeech, pp. 223–227 (2014)
6. Lin, Y.-L., Wei, G.: Speech emotion recognition based on HMM and SVM. In: Proceedings of the Fourth International Conference on Machine Learning and Cybernetics, Guangzhou, pp. 4898–4901, August 2005
7. Fundamentals of Telephone Communication Systems. Western Electric Company, p. 2.1 (1969)
8. Mathworks, eBook of Matlab and machine learning (2017). https://kr.mathworks.com/campaigns/products/offer/machine-learning-with-matlab.html
9. Kim, C.W., Yun, W.G.: Support vector machine and manufacturing application. In: Chaos Book (2015)
10. Vlachos, A.: Active learning with support vector machines. Master thesis, University of Edinburgh (2004)
11. Bae, M.J., Lee, S.: Digital Voice Analysis. Dongyoung Press (1987)
12. Ribiner, L.R., Schafer, R.W.: Theory and Applications of Digital Speech Processing. Pearson, Upper Saddle River (2011)
13. Park, H.W., Kim, M.S., Bae, M.-J.: Improving pitch detection through emphasized harmonics in time-domain. In: Kim, T., Ma, J., Fang, W., Zhang, Y., Cuzzocrea, A. (eds.) Computer Applications for Database, Education, and Ubiquitous Computing. Communications in Computer and Information Science (CCIS), vol. 352, pp. 184–189. Springer, Heidelberg (2012)
14. Suykens, J.A.K., Vandewalle, J.: Least squares support vector machine classifiers. Neural Process. Lett. **9**(3), 293–300 (1999)

15. Fung, G.M., Mangasarian, O.L.: Multicategory proximal support vector machine classifiers. Mach. Learn. **59**(2), 77–97 (2005)
16. Yoon, S.-H., Bae, M.-J.: Analyzing characteristics of natural seismic sounds and artificial seismic sounds by using spectrum gradient. J. Inst. Electron. Eng. Korea SP **46**(1), 79–86 (2009)
17. Huang, G.-B., et al.: Extreme learning machine for regression and multiclass classification. IEEE Trans. Syst. Man Cybern. Part B (Cybern.) **42**(2), 513–529 (2012)
18. Zheng, L., et al.: Integrating granger causality and vector auto-regression for traffic prediction of large-scale WLANs. KSII Trans. Internet Inf. Syst. (TIIS) **10**(1), 136–151 (2016)
19. Lee, D., Shin, D., Shin, D.: A Finger counting method for gesture recognition. J. Internet Comput. Serv. **17**(2), 29–37 (2016)
20. Su, C.-L.: Ear recognition by major axis and complex vector manipulation. KSII Trans. Internet Information Syst. **11**(3), 1650–1669 (2017)
21. Ishidaa, H., Oishib, Y., Moritac, K., Moriwakic, K., Nakajimab, T.Y.: Development of a support vector machine based cloud detection method for MODIS with the adjustability to various conditions. Remote Sens. Environ. **205**, 390–407 (2018)

A Study on the Factors Affecting a Switch to Centralized Cloud Computing for Local Government Information System

Lee Ki Bum[1], Lee Hyung Taek[1], Kim Jong Yoon[1],
and Gim Gwang Yong[2(✉)]

[1] Department of IT Policy and Management,
Soongsil University, Seoul, South Korea
kblee@klid.or.kr, htlee@innotium.com, kimjw@nice.co.kr
[2] Department of Business Administration, Soongsil University,
Seoul, South Korea
gygim@ssu.ac.kr

Abstract. This study examined the factors affecting the user's intention to switch seen in the stage of switching the administrative information system being operated by the local government to a centralized cloud computing and a research model was identified based on precedent researches in order to achieve the purpose of the study. As for the research model, it was established with regards to the switch intention of administrative information system by utilizing the migration theory's PPM model (Push-Pull-Mooring model) and the study aims to conduct an empirical test on it. As for the research model, properties that belong to the environmental factors of the place of origin and place of destination were set as independent variables in the migration theory's PPM model. As for the properties of Push Factors, public relations service, computational resource management cost and system quality were extracted and as for the properties of Pull Factors, economic feasibility, energy efficiency and scalability were extracted. As for the properties of Mooring Factors, switch cost and information security were extracted to verify the regulation effect. The aim was to analyze the effect on the switch intention through parameters of dissatisfaction and attractiveness of alternative with regards to the variables that belong to the respective factors that have been extracted. In order to conduct the empirical test on this study, an offline survey was conducted on 102 people in charge of computer system who operate information system hardware and people in charge of establishing policies in the local government. Empirical test was conducted on the survey data of 97 respondents excluding insincere responses using SPSS and PLS. As a result, public relations service and system quality affected dissatisfaction while the economic feasibility, energy efficiency and scalability affected the level of satisfaction of alternative. It was analyzed that both the attractiveness of alternative and dissatisfaction affected the switch intention and in terms of regulation effect, the information security had a significant impact.

Keywords: PPM model · Migration theory · Cloud computing
Centralized · Local government's administrative information system
Switch intention

© Springer Nature Switzerland AG 2019
R. Lee (Ed.): SNPD 2018, SCI 790, pp. 100–122, 2019.
https://doi.org/10.1007/978-3-319-98367-7_9

1 Introduction

The cost of IT resource management is increasing for local governments that are unable to cope with the rapidly changing IT environment and the inefficiency of IT resource utilization is growing. The local government is in charge of handling delegated work which involves handling the work delegated by the central government as well as the local work which involves handling the unique work of the local government. In particular, with regards to the delegated work, the local government has established and disseminated a standard system to resolve the informatization gap per local government. In addition, a standard work procedure is being applied and operated. The central and local government are constructing optimal operation environment for regional informatization by enhancing efficiency of administrative work by establishing organic, cooperative relationship and by providing customized administrative service.

The standard system distributed to the local government individually by the central government pursuant to the relevant acts has to provide unified functions based on mutual discussion between the government ministries but as it has been distributed with different purposes at different periods, it is difficult to respond flexibly to the changes in administrative environment such as reformation of local government's administrative district, enactment and revision of legal system and others.

Moreover, the specialized systems established and operated individually by the local governments are information systems that have been applied to process the unique work in accordance with the needs per local government so it is difficult to share and utilize information with the standard system that has been distributed by the central government. The need for a study on the switch to an integrated and cloud computing environment for information resources has been emphasized in order to implement joint utilization of information by the expansion of mutual information interconnectivity with other government ministries and local governments and to reduce the cost of IT resource maintenance and management following the operation of a computer room per local government and others.

There is a need to discuss about the composition of centralized cloud computing environment that could break-away from the IT operation based hollowization in local government following the construction of cloud system or data center and it would be able to contribute to the promotion of not only the reinforcement of local government but also the balanced regional development.

As a local government information system's operation environment has been established in individual institutes, it is difficult to promptly respond to the error in the standard system distributed by the central government and the need for establishing an integrated environment in low-cost, highly efficient structure has been raised to minimize the gap in informatization between regions such as financial independence and others. In consideration of the direction of local electronic government development, a realistic review on the cloud computer is required to share and open knowledge information and a technical and organizational diagnosis is required with regards to the switching of independent information system's operation environment to a centralized cloud computing environment for storage and utilization of local public data.

This study first aims to examine the administrative information system established and distributed as standard system by the central government to all local governments then analyze which characteristics out of those related to cloud computing service are factors that affect the switching of administrative information system to centralized cloud computing environment. In order to empirically analyze the effect and importance of various preceding variables that affect the switch intention of information system, a research model was established by utilizing the migration theory's PPM Model (Push-Pull-Mooring) and an empirical test was conducted using survey by setting hypothesis between each factor and switch intention.

2   Theoretical Background

2.1   Overview on Local Government Information System

Local government refers to a group constituted by residents of a particular region in order to run the administration and politics of the regional society in accordance with decision making independent from the central government. The local government is categorized into the metropolitan local authority and primary local authority. As for the local government, special local authorities, such as the Education Committee, exist other than the general local authorities. Metropolitan local authority includes special city, metropolitan city and province while the primary local authority includes city, district and autonomous district (district of special city and metropolitan city) [1].

The characteristics of local government differ from local administrative institutes under the central administrative institute in terms of status and trait. First, the local government is an incorporation in which rights and obligations separate from the country become the principal. Second, the local government is not just any incorporation but is a public corporation. In other words, it is a type of a public group created to process the local public works. Therefore, the local government is different from incorporated foundation or corporation. Third, the local government has autonomous right. As such, the local government is free from the control or interference of the central government and can independently process the work within the region. Forth, the local government has the characteristics of a regional group that has been founded so that the citizens of the region can independently process the work of the particular region. From this perspective, it is differentiated from public groups such as public association or Anstaltsperson [1].

The administrative information system used by the local government to process work is an overall information system in which approximately 300,000 government officials in 17 metropolitan local authority (city, province) and 226 primary local authority (city, county, district) can process civil service works and internal administration works online for the citizens and it supports transparent and efficient administrative work by organically interlinking the city, province, county, district, central government and related organizations.

The local government administrative information system is comprised of metropolitan system (city and province administration) and local system (city, country and district administration). The city and province administration system processes 22 administration works such as self-governing administration, city planning and others pursuant to 353 relevant acts in total (113 legal, 112 enforcement ordinances, 23 others). It registers and processes over 3,000 civil service works in a day on average. Although the work delegated to the city, country and district administration system differs, it processes 22 administration works such as civil defense, health and others pursuant to 462 relevant acts in total (141 legal, 139 enforcement ordinances, 140 enforcement regulations, 42 others). It registers 200,000 civil service works in a day on average. Sejong Metropolitan Autonomous City processes the work of both metropolitan and local authority so it has expanded the administrative information system for use. The administrative information system contributes to enhancing the efficiency of providing services for the convenience of citizens and of government official's work process by interlinking over 9,100 types of information for central government, public institute and local authority's information system such as Minwon 24, Haengbok-e and others (Fig. 1).

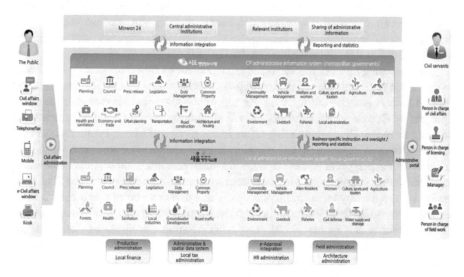

Fig. 1. Configuration of public administrative information system

The problem of the local authority's administrative information system is that it is old in terms of technical aspects. The application program was applied with the information system technology at the time of establishment (2002) so the functional coupling is low and the structure is complex, ultimately dropping the efficiency of maintenance. The hardware and commercial software (DBMS, WAS and others) operating the administrative information system is either discontinued or operates on old version in which the suspension of use is recommended due to the suspension of technical support and even if the computer resources such as server and others are replaced, new technology cannot be applied due to compatibility issues in some commercial software (web UI, reporting tool and others) and Java. In addition, the maintenance cost is continuously increasing.

2.2 Concept of Cloud Computing

In the era of the 4th industrial revolution, cloud computing has positioned itself as a new paradigm. It refers to the technology of a device with Internet connection being in charge of the interface (input and output) only and another computer connected to the Internet analyzing, processing and saving information. It is a form of providing service with regards to IT resources by utilizing information and communication network called the Internet. Cloud computing is a technology in which the desired computer resources are allocated at the desired time with regards to the large-scale, large-capacity processing requirements such as IT application service for use and it is equivalent to renting a car or using public transportation without buying the car.

Cloud service moves away from the method of users directly getting majority of resources for use such as server, storage, application program, network and others required for computing. Instead, it is a computing service in which part of computing resources is provided by using the network in order to enhance the efficiency and economic feasibility depending on the circumstances. This concept of cloud computing was introduced after a computer scientist, Professor John McCathy predicted in 1961 that "As the Time-Sharing technology evolves, computing power and application would be sold in Ability-type Business Model" [2].

The technology proposal with regards to cloud computing dates back to 2006 when an employee in Google started to contemplate. Nobody contemplated on the utilization method with regards to unused capacity out of the large-scale information resources operated by Google so the concept of cloud computing was proposed as a measure of utilizing them [3]. Cloud computing refers to the service of users paying the cost of IT resources such as application, OS security, storage and others based on how much is needed at a desired point of time through the Internet without professional technology and knowledge. Companies own and manage the IT resources and just like using electricity supplied by an electricity company and users can borrow and use all the IT resources conveniently at low cost through the Internet without large amount of money, time, workforce and venue [4, 5]. The NIST (National Institute of Standards and Technology) in USA defined cloud computing as a 'Model of borrowing and using from computing resources (software, storage, server, network) pool that can be constituted through the network based on how much is needed when needed and promptly providing them through service operators based on minimum management efforts [6].

Representative services provided by cloud computing are SaaS (Software as a Service), PaaS (Platform as a Service) IaaS (Infrastructure as a Service). SaaS which services application refers to the model of cloud service operator providing software at cloud computing server, users connecting to the Internet remotely to use the relevant software. PaaS is a model that provides the foundation for users to develop software and the cloud service operator supports compatibility provision service and service components based on PaaS. In addition, IaaS involves provided server infrastructure as a service and refers to a model which provides computing power or storage through cloud in the form of a service through the network [8].

Depending on the subject and scope of cloud service, it can also be classified into Private Cloud, Public Cloud and Hybrid Cloud. Private Cloud refers to services that are only provided to the internal organization members of a particular company (Internal

Cloud) while the Public Cloud refers to open-type of service that is open to anyone (External Cloud). Hybrid Cloud refers to being operated as Private Cloud for certain work and utilizing Public Cloud together for other works [8, 10]. The types of cloud services [11] examined beforehand has been organized in Table 1.

Table 1. Characteristics per cloud computing service types

Item	Main characteristics
Public Cloud	- A service that is implemented to be used by anyone and is provided to general users or large corporations and charged according to the usage amount - The infrastructure of a public cloud is owned by the company that sold the service
Private Cloud	- A service that provides cloud computing within a specific organization and implemented in a closed environment
Hybrid Cloud	- A mix of public and private clouds, keeping important materials in a private cloud, partially in the form of a public cloud - Bundled with standard technologies that enable the movement of data and applications, or consolidate two or more clouds
Community Cloud	- Services intended for common use by organizations and organizations in similar settings - Share a distributed relationship(Objective, Policy, Security Requirements and Convention)

To summarize, cloud computing is a next-generation Internet service in which resources are borrowed based on the desired amount at an affordable cost instead of owning and managing the IT resources directly and paying for the amount of actual use. As for cloud computing, the network infrastructure is extremely abstract from the user in relation to the server hardware so IT resources can be shared and unused resources can be utilized efficiently. In addition, it can reduce the cost of securing the bandwidth of circuit in preparation for situations in which service is focused temporarily. Furthermore, it can enhance the productivity and work efficiency by establishing permanent cooperation system. Main client companies maintain security and privacy with regards to their information while having the characteristics of multi-tenancy capable of using the cloud computing power. However, despite such usefulness, the concerns of users continuously exist from the perspective of stability, reliability and security [12].

As a preceding study related to cloud computing, Gupta et al. proposed 5 factors that would affect the cloud computing to be used by business communities. The most preferred factor is the convenience and easiness of users, followed by privacy and security and the third factor is cost reduction. As for reliability, it is ignored because the SMEs do not consider cloud to be reliable and as for fifth factor, SMEs stated that they do not wish to use cloud for cooperation and sharing and that they prefer the previous conventional method for cooperation and sharing with their interested parties [13].

Kim Tae Jin et al. wasn't able to analyze significant difference in terms of statistics out of 4 types of regional dispersion, centralized, dualization and maintenance of current system in the research conducted with regards to the local government's cloud service conducted on person in charge of informatization in local governments but it demonstrated that regional dispersion type was the most preferred model. The response that introducing cloud computing would be successful for materializing e-government was studied to be significant and it was analyzed that desk top virtualization was the highest as optional service [14].

In the study conducted by Jung Jae Ho et al. insecurity with regards materials and information stored in cloud, concerns on service stability, difficulty in service transition due to low standardization standard, other switching costs have been suggested as obstacle factors with regards to the expansion of cloud computing service and as a measure for improving the system for activation of cloud computing, user protection following business insolvency, user protection measures following temporary service error, solutions for resolving user security concerns have been suggested and as a strategy for technical response to activate cloud computing, measures to secure cloud computing service quality and measures to secure mutual operability of cloud computing have been suggested [15].

In the study conducted by Low et al. factors that affect the introduction of cloud in companies at the high-tech industry sector have been applied to TOE Framework. The 8 factors, relative advantage, corporate scale, complexity, CEO support, suitability, competition pressure, pressure of the other transaction party and technical preparation were verified [16].

Jeon Sae Ha et al. studied about the impact of cloud service's key characteristics on the public institute's service usage intention. UTAUT model and a model were set by observing the precedent study and based on it, the public sector's cloud computing acceptance behavior was described. As a result of analysis, all ubiquitous characteristics did not affect the expectation with regards to performance and the control variable of informatization leadership and innovative organization culture and others did not have a control effect in this research model. Moreover, external factors such as social impact and acceleration condition did not have an impact on the usage intention. This study suggests the considerations of applying cloud service in public sector. However, it is unfortunate that only part of the characteristics of cloud computing service were reflected and only the usefulness aspects of cloud computing were analyzed without considering the application cost or risk [12].

2.3   Migration Theory and PPM (Push-Pull-Mooring) Model

2.3.1   Migration Theory

Migration refers to a person or a group moving from one place to another. The migration of popular is explained by various social factors such as difference between two regions, in other words, difference in the employment opportunity, wage standard, standard of living and welfare, guarantee of human rights, education conditions and others [17].

Migration theory is related to people physically moving from a particular region to another region during a certain period of time [18]. When considering migration, the originating location and the destination location with permanence from a certain aspect must be clearly defined [19]. The migration of people can be defined as moving away from the currently unfavorable residence environment to an amicable environment [20]. Based on it, the Push-Pull model describing the various differences between the original location and the destination location was devised and it was explained by categorizing into the Push factor and Pull factor according to Bogue [21].

As for factors that restrict migration, the concept of 'Mooring' was suggested in 1992 to explain when the normative and psychosocial factors were acknowledged to be important [22] and since then, the Push-Pull model was added by Moon, expanding it as an original form of PPM model [23]. It can be said that the Mooring Factors which are the social and personal factors from microscopic perspective are added together with the Push Factors and Pull Factors that affect the decision of migration from macroscopic perspective.

2.3.2 PPM Model

PPM model is based on Push-Pull paradigm which was announced in the 1980s to explain migration theory [20, 24, 25, 29]. PPM model which is an important paradigm that explains the concept of migration provides useful, theoretical evidence describing the determinant of user's offline and offline service switching behavior from a particular system or service including cultural and geographical movement of human to a better service [26–28].

PPM model explains the migration factors with the push factor that makes people leave the current place of residence, pull factor that attracts them to a new place of residence as well as the interaction between Mooring factors which are factors that affect the migration between Push Factors and Pull Factors. Push Factor indicates negative factor that makes people leave the initial region to another region and the Pull Factor refers to the positive factor that attracts people to new destinations [24]. It has been pointed out that in the Push-Pull paradigm, the interaction of two factors are not enough to describe the migration of people so in order to explain the migration of people more effectively, the Mooring factor was applied additionally based on the result of studies so far [23, 24, 27]. Mooring factor also tends to directly affect the variable role of controlling between the Push and Pull factors or switch intention [19].

Push factor can be a factor that has a negative impact that makes people unsatisfied with the environment of the original place of residence and leave to a place of residence with a new environment [30]. Even with regards to the service switch behavior, Push factor, just like migration of people, is a concept that refers to the characteristics of existing service that influence the migration decision and it can be the driving force of searching for the directing point of a new service [23]. Generally, the Push factor is a factor that demonstrates negative images with regards to the current service such as low reliability, low quality and others but positive factors leading to the search for services of new alternative following the changes in social environment and individual's innovation can also be found.

Pull factor, as an extremely positive factor that attracts potential migrants, tend to create favorable impression in people [31]. In the study related to the switch behavior in the service field, the level of attractiveness of the alternative is most explained as a Pull factor [24]. A migrant waiting to move to a new place of residence will compare the various properties between the current place of residence and the new place of residence and even in the service switching research, it demonstrated that the intention of the user switching the service is positively connected to the perception of alternative [35]. The attraction of alternative affects even the products and services of a competitor company, ultimately having a positive impact on the switch intention of a consumer [32].

As the Push-Pull factor isn't sufficient to describe migration, the concept that has been added is the Mooring factor. Even if there is a strong correlation with the Push-Pull factor, in case the migration is not carried out due to a circumstance or condition, it contributes to the Mooring factor. Mooring factor can be an obstacle that interrupts the migration and the variable that belongs to it can include switching cost, attitude towards switching, social influence and others [24, 33]. Even the4 switching behavior of the service is considered to suppress the switch to a new service as a Mooring factor if the inertia of user with high dependency on the existing service is big or if the switching cost to a new service is high.

2.3.3   Precedent Study Related to PPM Model

As an element that categorizes service quality and satisfaction, the factor that affects the actual switch intention is said to be the service quality and not satisfaction [24]. However, as it has been measured fragmentarily from the perspective of service quality, the impact on the switch intention following a multi-lateral aspect related to service quality was not suggested.

Hou et al. conducted a study by applying the PPM model to a Massive Multiplayer Online Role Playing Game (MMORPG) and in this study, the insufficient awareness with regards to participants, low satisfaction and low joy were identified as Push factors, the attractiveness of alternative was identified as Pull factor and the low switching cost, high desire for diversity, weak social relationship and successful transfer and switching experience were identified as Mooring factors. In the research result, it was analyzed that low joy, low attractiveness of alternative and switching cost had a positive impact on the switch intention and the successful transfer and switching experience and the high desire for diversity had a negative impact on the switch intention [34].

The study conducted by Hsieh et al. constituted the factors based on PPM model with regards to the switch intention from blog to Facebook. As for the push factor, concern on writing posts and the weak connection between users seemed to have a positive impact on the switch intention. Compared to blog, Facebook's relative usefulness, convenience and joy had a positive impact on switching to Facebook. The past switching experience and switching cost had a negative impact on the switch intention, having a direct impact on reducing the switching behavior to Facebook and as a control effect, concern on writing posts and weak connection between users reduced the impact of switch intention [27].

Chang et al. suggested regret and dissatisfaction as the Push factors, attractiveness of alternative as a Pull factor and switching cost as Mooring factors as variables affecting the SNS switch intention. Push and Pull factors had a positive impact on the switch intention and the Mooring factor had a negative impact on the switch intention. In addition, the Mooring factor, as a control variable deteriorating the relationship between the switch intention and Push factors, and the Pull factor, as a control variable reinforces the relationship between the switch intention and Push factors, seemed to have a significant impact [26].

The service quality was said to have an impact on the customer satisfaction and the customer satisfaction was said to have an impact on purchase [36]. It can be assumed that if the level of customer satisfaction is high, the customer's possibility of switching drops so if the switch intention is high, the possibility of switching behavior is high and reversely, if the switch intention is low, the possibility of switch behavior is high [37].

Cho Hyeon et al. suggested the level of satisfaction, system quality, loyalty and service quality as Push factors, influence of acquaintances as Push factor and the installation cost, maintenance cost and embedding cost as Mooring factors for SNS users to describe the online service switching behavior. In their study, it was analyzed that the influence of acquaintance, installation cost and satisfaction had a positive impact on the switch intention [25].

3   Setting Research Model and Hypothesis

3.1   Research Model

In this study, PPM model which is a useful theory for explaining the switch determinant that has an impact when switching the currently used local government information system to a centralized cloud computing environment was applied. The environmental factors related to the switch intention from the originating location to the destination location were set as independent variables of Push, Pull and Mooring Factors and by applying dissatisfaction and attractiveness of alternative as parameters, a research model was established with high expectations that a research model with high effectiveness could be created.

Based on the existing precedent study, human service, management cost of computer resource and system quality were assigned as Push Factors of negative factors and as for the Pull factors of positive factors, economic feasibility, energy efficiency and scalability were assigned. Dissatisfaction was set as a parameter with regards to the Push Factors and the level of alternative was set as a parameter with regards to the Push factors in order to examine the relationship of parameter and the relationship of the dependent variable, a switch intention. Furthermore, the aim is to empirically verify how the Mooring Factors of switching cost and information security have an impact on the switch intention. The research study is as Fig. 2.

3.2   Setting Hypothesis

Hypothesis 1: Human service will have a positive impact on the dissatisfaction factor.
Hypothesis 2: Computer resource management cost will have a positive impact on the dissatisfaction factor.
Hypothesis 3: System quality will have positive impact on the dissatisfaction factor.
Hypothesis 4: Economic feasibility will have a positive impact on the attractiveness of alternative.
Hypothesis 5: Energy efficiency will have a positive impact on the attractiveness of alternative.

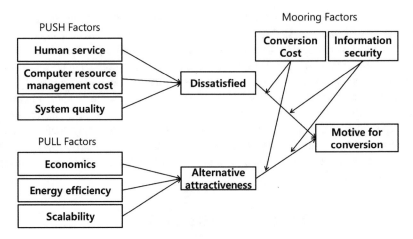

Fig. 2. Research model

Hypothesis 6: Scalability will have a positive impact on the attractiveness of alternative.

Hypothesis 7: Dissatisfaction will have a positive impact on the switch intention.

Hypothesis 8: Attractiveness of alternative will have a positive impact on the switch intention.

Hypothesis 9: Switching cost will control the relationship between the attractiveness of alternative and the switch intention.

Hypothesis 10: Switching cost will control the relationship between dissatisfaction and the switch intention.

Hypothesis 11: Information security will control the relationship between the attractiveness of alternative and the switch intention.

Hypothesis 12: Information security will control the relationship between dissatisfaction and the switch intention.

4   Empirical Testing

4.1   Material Collection and Analysis Method

This study established a research model and set a hypothesis based on precedent research in order to analyze the switch intention to a centralized cloud computing for the local government information system. Currently, an Information System Master Plan (ISMP) is expected to be established for the local government information system that is outdated and additionally, measures for establishing information resources are expected to be prepared based on efficient operation of the administrative information system. For an effective study, offline survey was conducted on the people in charge of information system in the local government and 102 responses were collected in total. Excluding 5 insincere responses, 97 survey data was used for final analysis. As for statistical analysis tool, SPSS and smart PLS were utilized.

4.2 Characteristics of Sample

Frequency analysis was conducted to determine the general characteristics of the survey respondents with the sample as the standard and the result is as Table 2. The number of respondents of the analysis sample were 97 people with 68 males (70.1%) and 29 females (29.9%). As per institute, it was distributed in the order of 73 people (74.2%) from primary local authority, 17 people (17.5%) from the central government and 8 people from the metropolitan local authority and as for the academic background, 3 people were high school graduates (3.1%), 88 people were university graduates (90.7%) and 4 people were graduate school graduates (4.1%).

Table 2. Demographic characteristics of sample

Item		Frequency (people)	Composition ratio (%)
Institute classification	Primary local authority (city, county, district)	72	74.2
	Metropolitan local authority (city, province)	8	8.3
	Central government	17	17.5
Gender	Male	68	70.1
	Female	29	29.9
Age	20 s	2	2.0
	30 s	29	29.9
	40 s	53	54.6
	50 s	13	13.5
Academic background	High school graduate and below	3	3.1
	University student	2	2.1
	University graduate	88	90.7
	Graduate school graduate	4	4.1
Class	Secretary (Class 4)	0	0.0
	Deputy director (Class 5)	3	3.1
	Class 6	30	30.1
	Class 7	48	49.4
	Class 8	12	12.3
	Class 9	4	4.1
Series	Computing position	92	94.8
	Administrative position	5	5.2
	Communications position	0	0.0
Years of service	Above 1 year, below 4 years	7	7.2
	Above 4 years, below 8 years	18	18.6
	Above 8 years, below 10 years	7	7.2
	Above 10 years, below 15 years	28	28.9
	Above 15 years	37	38.1

As for series, computing position took up the majority with 92 people (94.8%) and as for the class, it was in the order of 48 people in class 7 (49.4%), 30 people in class 6 (30.1%) and 12 people in class 8 (12.3%). As for the years of service, 72 people had worked for more than 8 years (74.2) and 25 people had worked less than 8 years (25.8%). Looking at the age, series and position distribution, they were in charge of practical work related to informatization at the front in the local government and were positions that suggested the most opinions on the construction and application of information system.

4.3   Exploratory Factor Analysis and Reliability Analysis

Validity analysis and reliability analysis were conducted in order to verify the suitability of the research model. Cronbach's Alpha coefficient was used in the reliability analysis in this study and the coefficient threshold was set as 0.7 and in case the value is above 0.7, it was judged that there was no problem in the reliability of the subject variable [38].

For verification, the extent of correlation embedded in the observation variables was explored through EFA (Exploratory Factor Analysis) in order to analyze the structure between factors. As for the factor analysis, it was conducted by setting the unique standard of the factors as 1.0 based on Varimax orthogonal rotation method and cases in which the factor loading was above 0.5 was judged to be significant [39].

As for the result of exploratory factor analysis and reliability analysis, energy efficiency, information security, switch intention, human service, switching cost, dissatisfaction, scalability and attractiveness of alternative were tied well with above 0.5 but as the computer resource management cost item 4, 5 and economic feasibility item 4, 5 didn't bind well so they were excluded from the analysis.

However, in general, the factors were tied well per factor and since Cronbach-α is $0.829 \sim 0.952$, being higher than 0.7, reliability and validity are deemed to have been secured. The analysis result is as is Table 3.

Table 3. Result of exploratory factor analysis and reliability analysis

	Composition factor											Cronbach-α
	1	2	3	4	5	6	7	8	9	10	11	
System quality 3	.856											0.925
System quality 4	.823											
System quality 2	.799											
System quality 5	.786											
System quality 1	.779											
System quality 7	.720											
System quality 6	.684											
Energy efficiency 4		.890										0.943
Energy efficiency 3		.850										
Energy efficiency 2		.846										
Energy efficiency 1		.804										
Energy efficiency 5		.796										

(*continued*)

Table 3. (*continued*)

	Composition factor											Cronbach-α
	1	2	3	4	5	6	7	8	9	10	11	
Information security 6			.890									0.928
Information security 5			.871									
Information security 4			.831									
Information security 1			.824									
Information security 3			.792									
Information security 6			.754									
Switch intention 2				.905								0.952
Switch intention 4				.867								
Switch intention 1				.867								
Switch intention 3				.866								
Human service 5					.898							0.925
Human service 2					.880							
Human service 3					.834							
Human service 1					.825							
Human service 4					.798							
Switching cost3						.844						0.863
Switching cost 2						,802						
Switching cost 1						.794						
Switching cost 5						.724						
Switching cost 4						.712						
Dissatisfaction 1							.830					0.941
Dissatisfaction 3							.824					
Dissatisfaction 2							.811					
Dissatisfaction 4							.811					
Computer resource management cost 2								.897				0.87
Computer resource management cost 3								.849				
Computer resource management cost 1								.791				
Computer resource management cost 6								.694				
Scalability 2									.784			0.829
Scalability 3									.745			
Scalability 1									.632			
Scalability 4									.554			
Economics 2										.729		0.876
Economics 1										.702		
Economics 5										.635		
Alternative attractiveness 3											.669	0.931
Alternative attractiveness 2											.667	
Alternative attractiveness 1											.581	

4.4 Confirmatory Factor Analysis

4.4.1 Model Compatibility Analysis

As a compatibility index with regards to the entire structure model, there is Stone-Geisser Q2 test statistics, a cross-verified Redundancy index and this shows compatibility of a structure model as a statistics estimated value and its value must be a positive number [40, 41]. According to Cohen [42], the degree of effect of R2 value is classified as high (more than 0.26, middle (0.13 ~ 0.26) and low (0.02 ~ 0.13). The entire compatibility has been obtained by multiplying the average value of Communality with the average value of R2 then finding the square root [41]. The size of this compatibility is classified as high (0.36 ~), middle (0.26 ~ 0.36) and low (0.1 ~ 0.25).

As for the compatibility of evaluation model of this study, the evaluation result exceeds all standards so the research model has explanation power. The result of model compatibility analysis is as Table 4.

Table 4. Result of model compatibility analysis

	Standard		Result			
Redundancy	≧0 (positive number)		Attractiveness	0.128609		
			Dissatisfaction	0.329094		
			Switch intention	0.330838		
Model fit(R2)	0.26 ~	high	Attractiveness	0.465118		
	0.13 ~ 0.26	middle	Dissatisfaction	0.397581		
	0.02 ~ 0.13	low	Switch intention	0.397764		
Total model fit (R2medium X Communality medium) Square root	0.36 ~	High	0.567856		R2	0.420154
	0.25 ~ 0.3	Middle			Communality	0.76748
	0.1 ~ 0.25	low			Total fit	0.567856

4.4.2 Reliability Analysis

As it has been confirmed that the model compatibility had explanation power, the reliability analysis of confirmatory factor analysis was conducted.

By conducting the reliability analysis of confirmatory factor analysis, the AVE (Average Variance Extra) of all variables was above 0.5 and the complex reliability was above 0.75, indicating that it was reliable [43]. The result of the reliability analysis is as Table 5 below.

Table 5. Result of reliability analysis

	AVE	Composite Reliability	R Square	Cronbachs Alpha	Communality	Redundancy
Economic feasibility	0.805907	0.925604		0.879112	0.805907	
Attractiveness of alternative	0.879133	0.956178	0.465118	0.931241	0.879133	0.128609
Dissatisfaction	0.850901	0.958016	0.397581	0.941525	0.850901	0.329094
System quality	0.700681	0.941994		0.927302	0.700681	
Energy efficiency	0.818699	0.957541		0.944327	0.818699	
Human service	0.772537	0.944314		0.926266	0.772537	
Computer resource management cost	0.541461	0.767166		0.887503	0.541466	
Switch intention	0.875332	0.965611	0.397764	0.952451	0.875332	0.330838
Scalability	0.662664	0.886347		0.829717	0.662664	

4.4.3 Discriminant Validity Analysis

As it has been confirmed that the model compatibility had explanation power, the reliability analysis of confirmatory factor analysis was conducted and in order to check the validity, discriminant validity analysis was conducted.

As for the result of this study's discriminant validity analysis, the correlation with the highest value in the correlation matrix of potential variable is economic feasibility and energy efficiency, with 0.662467. The square root of the correlation coefficient, in other words, the coefficient of determinant is 0.438862 (0.662467 × 0.662467) so the AVE value calculated in all potential variables was higher than the coefficient of determinant, ultimately securing discriminant validity. The result of discriminant validity analysis is as Table 6.

4.5 Hypothesis Testing

4.5.1 Result of Hypothesis Testing

PLS program was used to verify the research hypothesis and the result of statistical analysis is as Table 7.

Based on Table 7, if the characteristic factors of this research are summarized per hypothesis, it is as Table 8.

4.5.2 Analysis of Hypothesis Verification Result

In this study, it was analyzed that out of the Push Factors, human service and system quality had significant impact on dissatisfaction but the computer resource management cost didn't have a significant impact on dissatisfaction. It was analyzed that the Pull Factors of economic feasibility, energy efficiency and scalability all had a significant

Table 6. Result of discriminant validity analysis

	Economic feasibility	Attractiveness of alternative	Dissatisfaction	Quality	Energy efficiency	Human service	Computer resource management cost	Switch intention	Scalability
Economic feasibility	0.805907								
Attractiveness of alternative	0.502695	0.879133							
Dissatisfaction	-0.049489	-0.000787	0.850901						
System quality	-0.139966	-0.146014	0.623738	0.700681					
Energy efficiency	0.662467	0.520213	-0.018733	-0.115632	0.818699				
Human service	-0.204635	-0.003846	0.291704	0.33972	-0.234998	0.772537			
Computer resource management cost	0.290679	0.097993	0.075646	0.063144	0.063135	0.019577	0.541461		
Switch intention	0.389682	0.615628	0.136507	-0.069764	0.399712	0.083529	0.14869	0.875332	
Scalability	0.427598	0.608409	0.068594	-0.092447	0.442915	-0.034023	0.328774	0.435979	0.662664

Table 7. Result of statistical analysis of hypothesis verification

	Original Sample (O)	Sample Mean (M)	Standard Deviation (STDEV)	Standard Error (STERR)	T Statistics (|O/STERR|)	
Economics → Attractiveness of alternative	0.176287	0.176666	0.062367	0.062367	2.826605	Selected
Attractiveness of alternative → Switch intention	0.615726	0.614527	0.055154	0.055154	11.163702	Selected
Dissatisfaction → Switch intention	0.136784	0.135361	0.055087	0.055087	2.483059	Selected
System quality → Dissatisfaction	0.555469	0.563249	0.050068	0.050068	11.09423	Selected
Energy efficiency → Attractiveness of alternative	0.208181	0.205758	0.082575	0.082575	2.521106	Selected
Human service → Dissatisfaction	0.12984	0.131882	0.051257	0.051257	2.533131	Selected
Computer resource management cost → Dissatisfaction	0.151327	0.120807	0.149595	0.149595	1.011577	Dismissed
Scalability → Attractiveness of alternative	0.440822	0.440184	0.093336	0.093336	4.72297	Selected

Table 8. Result of hypothesis verification

Item	Hypothesis details	Result
Hypothesis 1	Human service → Dissatisfaction	Selected
Hypothesis 2	Computer resource management cost → Dissatisfaction	Dismissed
Hypothesis 3	System quality → Dissatisfaction	Selected
Hypothesis 4	Economics → Attractiveness of alternative	Selected
Hypothesis 5	Energy efficiency → Attractiveness of alternative	Selected
Hypothesis 6	Scalability → Attractiveness of alternative	Selected
Hypothesis 7	Dissatisfaction → Switch intention	Selected
Hypothesis 8	Attractiveness of alternative → Switch intention	Selected

impact on the attractiveness of alternative. In addition, the result of hypothesis verification indicated that dissatisfaction on switch intention and the attractiveness of alternative both had a significant impact.

4.5.3   Verification of Control Effect

An analysis on control effect was conducted with regards to the switching cost and information security variable in order to analyze the difference between groups following the characteristics of survey respondents and it was separated into group with high switching cost and group with low switching cost with regards to the result of survey by respondents for analysis. It was confirmed whether there was a difference in the two groups and multiple cluster analysis was conducted for it.

As a result of verifying the control effect, as all the P value was above 0.1, there was no difference between the group of control variables so it was analyzed that there was no control effect but as for information security, when the difference between the two groups were compared, the P value was below 0.1 among dissatisfaction and switch intention, energy efficiency and attractiveness of alternative, human service and dissatisfaction, scalability and attractiveness of alternative, indicating a difference (Table 9).

Table 9. Result of verifying the control effect of information security control variable

Hypothesis	Standard coefficient	T value	Hypothesis verification	Analysis of difference in path (P value)	
Economics → Attractiveness of alternative	0.216054	3.165286	Selected	0.124	Not different
	0.089041	1.774851	Selected		
Attractiveness of alternative → Switch intention	0.558052	9.083668	Selected	0.248	Not different
	0.649419	12.681989	Selected		
Dissatisfaction → Switch intention	0.298907	5.792686	Selected	0	Different
	-0.000344	0.005706	Dismissed		
System quality → Dissatisfaction	0.619966	9.969089	Selected	0.298	Not different
	0.542209	11.954449	Selected		

(continued)

Table 9. (*continued*)

Hypothesis	Standard coefficient	T value	Hypothesis verification	Analysis of difference in path (P value)	
Energy efficiency → Attractiveness of alternative	0.084254	1.208988	Dismissed	0.094	Different
	0.26622	3.252015	Selected		
Human service → Dissatisfaction	0.038392	0.509517	Dismissed	0.054	Different
	0.197359	4.553513	Selected		
Computer resource management cost → Dissatisfaction	0.094898	1.160973	Dismissed	0.499	Not different
	0.29892	1.161531	Dismissed		
Computer resource management cost → Dissatisfaction	0.684977	16.040882	Selected	0	Different
	0.14102	1.416554	Dismissed		

5 Conclusion

5.1 Research Result and Implication

In this study, an empirical study was conducted with regards to factors that affect the switch to a centralized cloud environment for the local government information system. The characteristics of cloud computing were analyzed and based on the precedent studies related to the switch of information system, a research model was established. The verification of the research model's hypothesis was conducted through empirical analysis and based on it, the following results were identified.

First, in the hypothesis that Push Factors, as negative factors with regards to the current system, have an impact on dissatisfaction, human service and system quality were selected and the computer resource management cost was dismissed. In the many current researches, the system quality and maintenance workforce's human service are closely related to satisfaction with regards to the current service and it is judged that due to the poor timing of responding to error following the outdated current system and long-distance maintenance, there is a director relationship with dissatisfaction. Moreover, the hypothesis with regards to computer resource management cost was dismissed. Every local government is reducing the maintenance cost by efficiently operating computer resources through integrated, abolition and virtualization of information system so it is interpreted that the computer resource management cost does not affect dissatisfaction.

Second, in the hypothesis that Pull Factors, as positive factors, affect the attractiveness of alternative, all 3 factors of economic feasibility, energy efficiency and scalability were selected. It can be said there is positive impact on the people in charge of computer at local government in the sense that cost reduction can be expected when switching to cloud computing environment. Energy efficiency, which is an environment-friendly factor that can reduce space and power consumption, is being managed by the Ministry of Environment with regards to the annual energy use and greenhouse gas emission even in the central government and local government pursuant to the Framework Act on Low Carbon, Green Growth. Following such, with the introduction of green computing environment of the computer room, it seems that energy efficiency is expected to increase with regards to the use of electricity. As for scalability, there is no big change in the operation method when the current system is switched to a centralized cloud computing and depending on the system usage amount, the resources can be allocated and recovered variably so it can be said that it would have a positive impact on the people in charge of operating the local government system by securing the operation stability of computer resources.

Third, the hypothesis with regards to system dissatisfaction and the hypothesis of attractiveness of alternative were selected as factors that affected the switch intention. The dissatisfaction factor with regards to the administrative information system currently being operated and the advantage of centralized cloud computing have been indicated as important determinants in the system switch intention.

Forth, an analysis on control effect was conducted with regards to switching cost and information security. As a result of analysis on the control effect, switching cost was not significant and with regards to information security, significant control effect was identified. The current local government information system is in dire need of system advancement as it is outdated so it is acknowledged that the switching cost with regards to such can be replaced by advancement budget. Thus, the control effect of the switching cost is analyzed to be insignificant. As for information security, data is integrated and operated in a single system, as a centralized cloud computing environment so as a concern for information security such as security vulnerability, it is interpreted that significant control effect is seen.

The implication of this study is that it can become practical reference materials with regards to the switch to a centralized cloud computing environment for the national standard distribution system that is being operated by the local government by utilizing the computer resource together. The cloud characteristics were analyzed and based on the academic materials, a research model was set and through empirical analysis, factors that affect the intention to switch the local government information system to a centralized cloud computing environment were verified.

5.2   Limitations of the Research and Future Research Direction

First, in this study, data was collected from the person in charge of computer in the local government and people in charge of policy in the central government in order to verify the hypothesis but considering the fact that there are approximately 300,000 government officers in the local government, there is a need to secure more samples. Furthermore, it would be better to conduct a study on people in charge of work at the local government and private entrepreneurs in charge of the maintenance of administrative information system who have more in-depth understanding on the current system.

Second, there is a need to study other characteristics of cloud service that is not proposed as independent variables in this research model. In the follow-up researches, it is expected that a model design with high explanation power could be achieved by finding independent variables that are more suitable to the proposed model and by eliminating certain variables from the model.

References

1. Lee, J.S.: The Public Administrative Dictionary. Daeyoung Moonhwasa Publisher, Seoul (2009)
2. Kim, T.G., Kim, T.H., Yang, J.Y., Yang, H.D.: Cloud Computing. Han Kyoung Sa (2011)
3. Kim, H.J.: Era of cloud computing: Cloud computing in which information can be pulled out and used just like turning on the tap water anywhere. Tech Future **51**, 14–17 (2008)
4. Lee, J.Y.: Characteristics of cloud computing and current status of service provided per service provider. In: Korea Information Society Development Institute, Broadcasting and Telecommunication Policy, vol. 22, no. 6 (2010)
5. Armbrust, M., Fox. A., Griffith, R., Joseph, A. D., Katz, R., Konwinski, A., Lee, G., Patterson, D., Rabkin, A., Stoica, I., Zaharia, M.: Above the clouds: a Berkeley view of cloud computing. UC Berkeley TR (2009)
6. Mell, P., Grance, T.: The NIST Definition of Cloud Computing (2011)
7. Lim, J.S., Oh, J.I.: A study on the effect of the introduction characteristics of cloud computing services on the performance expectancy and the intention to use- focusing on the innovation diffusion theory. Asia Pac. J. Inf. Syst. **22**(3), 99–124 (2012)
8. Min, O.K., Kim, H.Y., Nam, G.H.: Technology trend of cloud computing. Electron. Telecommun. Trends **24**(4), 1–13 (2009)
9. Ministry of Government Administration and Home Affairs: A study on the feasibility of implementing cloud for local electronic government (2014)
10. Park, S.C., Kwon, S.J.: A study on factors affecting intention to switch for using cloud computing. Korea IT Serv. Acad. J. **10**(3), 149–166 (2011)
11. Jung, W.J.: Legal problems for activating cloud computing–issues related to personal information protection. Korea Inf. Soc. Dev. **26**(20) (2014)
12. Jeon, S.H., Park, N.R., Lee J.J.: A study on the factors affecting the intention to adopt cloud computing service **10**(2), 97–112 (2011)
13. Gupta, P., Seetharaman, A., John, R.R.: The usage and adoption of cloud computing by small and medium businesses. Int. J. Inf. Manage. **33**(5), 861–874 (2013)
14. Kim, T.J., Hwang, S.S., Seo, S.H., Kim, D.H.: Designing cloud computing system for local governments: In pursuit of an optimal model utilizing case study and feasibility study. Korea Assoc. Reg. Inf. Soc. **18**(1), 85–107 (2015)
15. Jung, J.H., Gi, E.H.: Cloud computing, market opportunity and policy direction. Korea Inst. Commun. Inf. Sci. Acad. Conf. J. **2009**(6), 2041–2042 (2009)
16. Low, C., Chen, Y.: Understanding the determinants of cloud computing adoption. Ind. Manage. Data Syst. **111**(7), 1006–1023 (2011)
17. Lee, S.R.: Migration and population: Explanation from demographic perspective. In: IOM Migration Research & Training Centre's working Paper Series (2011)
18. Clark, W.A.V.: Human Migration. Sage Publications, Beverly Hills (1986)
19. Lee, E.S.: A theory of migration. Demography **3**(1), 47–57 (1966)
20. Ravenstein, E.G.: The laws of migration. J. Stat. Soc. Lond. **48**(2), 167–235 (1885)
21. Bogue, D.J.: Principles of Demography. John Wiley and Sons, New York (1969)
22. Longino, C.F.: The Forest and the Trees: Micro-Level Consideration in the Study of Geography Mobility in Old Age. Billhaven, London (1992)
23. Moon, B.: Paradigms in migration research: exploring "moorings" as a schema. Prog. Hum. Geogr. **19**(4), 504–524 (1995)
24. Bansal, H.S., Taylor, S.F., James, S.Y.: Migrating to new service providers: toward a unifying framework of consumers' switching behaviors. J. Acad. Mark. Sci. **33**(1), 96–115 (2005)

25. Cho, H., Park, S.S., Lee, G.B.: Cyber migration: an empirical study on online service switching. Telecommun. Rev. **23**(4), 472–486 (2013)
26. Chang, I.C., Liu, C.C., Chen, K.: The push, pull and mooring effects in virtual migration for social networking sites. Inf. Syst. J. **2**(4), 332–347 (2014)
27. Hsieh, J.K., Hsieh, Y.C., Chiu, H.C., Feng, Y.C.: Post-adoption switching behavior for online service substitutes: a perspective of the push-pull-mooring framework. Comput. Hum. Behav. **28**(5), 1912–1920 (2012)
28. Zengyan, C., Yinping, Y., Lim, J.: Cyber migration: an empirical investigation on factors that affect users' switch intentions in social networking sites. In: System Sciences, HICSS 2009. 42nd Hawaii International Conference on IEEE, pp. 1–11 (2009)
29. Lewis, G.J.: Human Migration: A Geographical Perspective. Croom Helm, London and Canberra (1982)
30. Stimson, R.J., Minnery, J.: Why people move to the 'sun-belt': a case study of long-distance migration to the Gold Coast, Australia. Urban Stud. **35**(2), 193–214 (1998)
31. Dorigo, G., Tobler, W.: Push-pull migration laws. Ann. Assoc. Am. Geogr. **73**(1), 1–17 (1983)
32. Jones, M.A., Mothersbaugh, D.L., Beatty, S.E.: Switching barriers and repurchase intentions in services. J. Retail. **76**(2), 259–277 (2000)
33. Anderson, E.W.: Cross-category variation in customer satisfaction and retention. Mark. Lett. **5**(1), 19–30 (1994)
34. Hou, A.C.Y., Chern, C.C., Chen, H.G., Chen, Y.C.: Migrating to a new virtual world: Exploring MMORPG switching through human migration theory. Comput. Hum. Behav. **27** (5), 1892–1903 (2011)
35. Zhang, K.Z., Cheung, C.M., Lee, M.K.: Online service switching behavior: the case of blog service providers. J. Electron. Commer. Res. **13**(3), 184–197 (2012)
36. Lapierre, J., Filiatrault, P., Chebat, J.C.: Value strategy rather than quality strategy: a case of business-to-business professional services. J. Bus. Res. **45**(2), 235–246 (1999)
37. Bitner, M.J.: Evaluating service encounters: the effects of physical surroundings and employee responses. J. Mark. **54**(2), 69–82 (1990)
38. Kim, G.S.: AMOS 18.0 Structural Equation Model. Hannarae Academy, Seoul (2010)
39. Young, L.H.: Professor Lee Hoon Young's Data Analysis Using SPSS, 2nd edn. Cheongram (2013)
40. Chin, W.W.: Issues and opinion on structural equation modeling. MIS Q. **22**(1), 1–8 (1998)
41. Tenenhaus, M., Esposito, V.V., Chatelin, Y.M., Lauro, C.: PLS path modeling. Comput. Stat. Data Anal. **48**(1), 159–205 (2005)
42. Cohen, J.: Statistical Power Analysis for the Behavioral Science, 2nd edn. Lawrence Erallbaum, Hillsdale (1988)
43. Fornell, C., Larcker, D.F.: Evaluating structural equation models with unobservable variables and measurement error. J. Mark. Res. **18**(1), 39–50 (1981)

Customer Prediction on Parking Logs Using Recurrent Neural Network

Liaq Mudassar and Yungcheol Byun[✉]

Jeju National University, Jeju, South Korea
mudassar192@hotmail.com, ycb@jejunu.ac.kr

Abstract. With each passing day, neural networks are being explored extensively, complimented by an equally swift rise in the usage of their applications, which, in turn are also seeing a rising trend. A neural network has two major sub casts; convolutional and recurrent neural networks. In this paper, we propose customer influx based on parking data as we attempt to predict customer flow of a large grocery store using its parking logs. Our aim is to solve the problem of vehicle parking by predicting the traffic flow within a specific traffic parking lot, with respect to time, using RNNs. We present a RNN which is trained on car parking logs. Using this information and traffic patterns of the days already past, our proposed RNN predicts traffic pattern for the following day and approaching week.

Keywords: Neural network · Convolutional · Recurrent · Prediction
Traffic patterns · Long short-term memory · Vanilla

1 Introduction

Parking is a serious concern when it comes to cities, especially big cities, according to Donald C. Shoup [1,12]. Finding a suitable parking space before one can go shopping, is an arduous task during rush hours and especially during holidays. According to Donald C. Shoup [1], a hefty 34% of the cars in a congested area will cruise for a parking space during peak hours. May it be Paris, Tokyo, Singapore or Seoul, it is a stiff task finding a parking space, especially if you have decided to go grocery shopping. According to Caicedo, Robuste, and Lopez-Pita [2] drivers familiar with the lay of the land are usually 45% more successful in finding parking spaces than those without the general know-how of the place. Generally, the parking spaces around grocery stores is at its full capacity, making you wait for parking to become available or making it so, that you would be forced to choose another store. This is a quandary for a grocery store owner, for whom, during peak hours, due to limited parking space, customer influx would remain constant. When the peak hours are over, although the parking space would be scarcely occupied, so will be his store.

Many researchers have tried to solve this problem with various techniques. In the context of mathematicians and engineers, it is a credit assignment problem

© Springer Nature Switzerland AG 2019
R. Lee (Ed.): SNPD 2018, SCI 790, pp. 123–136, 2019.
https://doi.org/10.1007/978-3-319-98367-7_10

of time series with prediction intended [11]. For mathematicians, the problem of credit assignment is rudimentary. It is a problem of assigning correct weights to different factors in a scenario, to get the desired output. Machine learning via neural networks is essentially an application of credit assignment problem, just on a larger scale and at different levels. In deep learning we learn credits across many different layers, to tune output to desired results.

Gradient decent with back propagation (which is a very common method used in NNs now a days) has been around since the early 60's and 70's. Its integration into the neural network started during 1981 [11]. Limelight for machine learning diminished in late 80's and 90's, when it was outperformed by other methods, because of its limitations to architecture and the processing power of computers. It again got attention post 2005 when computing power became significant enough and gained popularity when it was able to win the visual recognition competition in 2009 [11]. Since then, machine learning has been applied to most of the fields and it has proven to be state of the art at visual recognition and pattern recognition among many others.

Machine learning or deep learning works on two basic architectures. First type is called convolutional neural networks or simply CNNs. These networks are great in most of the scenarios like picture recognition, where data set is independent of any time or order constraints. In any scenario, where there are time or order constraints like video recordings, audio or text, CNNs fail to capture these dependencies. Recurrent neural networks or RNNs are the second type of neural networks which tackle this problem and are good in scenarios where there are temporal or order dependencies. They usually contain a memory component which lets them pass the selective knowledge down the time line.

Both CNN and RNN fall under artificial neural networks (ANN) or simply neural networks. Figure 1 elaborates ANN in more detail. As shown if Fig. 1 an NN usually contain 3 different type of layers. First is input layers via which input data is supplied to neural network. Next it contains X hidden layers.

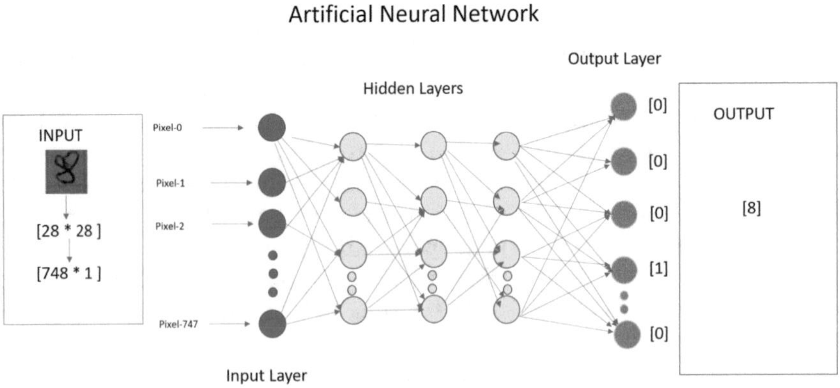

Fig. 1. Artificial neural network: overview

These hidden layers are the ones which extract abstractions out of data where each layer builds on abstractions provided to it from it previous layers. This X can very from 1 to multiple thousands depending upon complexity of problem. These hidden layers learn abstractions by learning weight matrix from the input data and these weight matrix are the core components used to represent learning by these layers. The final component in NN architecture is output layer which may contain 1 to many neutrons. This layer is responsible to produce output which can be range from a numerical value (in case of RNNs) to a classification label (in case of CNN's). All these layers are composed of neurons. Working of this artificial neuron is rather simple. Typically a neuron works in very simple manner. It takes all the inputs attached to it, multiplies it with its weight matrix and then applies a function on it to get its output. This function can be a simple aggregation to complex non-linear function. Following equation shows very thing in mathematical form.

$$a = \sum_{i=1}^{n} (x_i.w_i). \tag{1}$$

where

$x_0 =$ Initial bais (usually defaults to 1).
$x_1..x_n =$ inputs to neuron.
$w_1..w_n =$ weights associated to each input.

From the prospective of grocery store owner, it would be beneficial to know what customer trend would be tomorrow, day after tomorrow, or the following week. With this information, the owner can greatly benefit; by cutting costs in multiple perspectives based on customer forecast. A very simple example would be managing its employees. In case there is forecast of low customer response, owner can reduce the staff for next day and similarly can add extra workers in case of high customer arrival forecast.

In this paper we try to predict the parking trend for above mentioned situations and hence predict customer trend for the coming days. We use parking dataset as input and predict average number of cars for the days to come. In terms of a store parking lot, this directly translates to the number of customers arriving to the store. We use machine learning approach to try and solve this problem. We utilize multiple variations of RNN i.e. Vanilla and LSTM, on these historical parking logs to predict what would be the pattern for the upcoming days.

This Paper is divided into 5 sections. Section 1 is Introduction, Sect. 2 is related work, Sect. 3 is Proposed Methodology, Sect. 4 is Experimental Results and Sect. 5 includes Conclusion.

2   Related Work

The problem we are working on is customer prediction using parking data analysis. There hasn't been a lot of work done in terms of parking prediction using machine learning. The next most pertinent work done, is in the field of traffic prediction. In [3] Manoel Castro-Neto and Young-Seon Jeong have used supervised learning to predict short term traffic flow. They have used support vector machine with statistical regression (OL-SVR) to predict the flow for typical and atypical conditions. In addition, they used PeMS (California Highway Performance Measuring System) data and compare their result with Gaussian maximum likelihood, achieving equal or better performance in common scenarios. Haiyang Yu, Zhihai Wu in [4] predicted traffic using spatiotemporal RNN's with LSTM. They converted traffic features into images and then trained RNN on those images to predict the traffic flow. They didn't provide any comparison with related approaches. In [5] Sherif Ishak has tried to optimize traffic prediction using four different architectures. He has compared neural networks to LSTMs, principle component analysis co-adaptive neuro-fuzzy inference system (CANFIS) and modular networks. He has used videos varying from 1–10 mins length. He concluded that no single algorithm is best for all circumstances, as different algorithms perform better with varying circumstances. Rose Yu, Stephan Zheng in [6] have been doing long term for casting, using tensor trained RNNs. They have been using higher order generalization of RNN's and have captured temporal long-term dependencies. Their RNN has been equally good on both linear and non-linear environments. Their work has depicted better performance than both LSTM and vanilla RNNs.

These works mostly focus on traffic on the roads. i.e. what is the traffic intensity and pattern on the road. In our scenario it is a different use case, as we are considering data for a specific parking lot, potentially of a big shopping mart, that is concerned with its customer's patterns and usages. These works are valid in a way that they are using the same neural network that we are using in our work and they are also addressing traffic density, however, we are focusing on parking. Our focus in on static parking and prediction of customer's influx with respect to it. This makes this a different problem than what above work addresses, in terms of goal and execution.

There has been relatively less work done in the field of customer predictions using machine learning. In [7] Philip and Thomoson worked on churn rate (customer leaving) prediction. They focused on customer features which are not specific to any firm and are independent. Like Philip and Thomoson most of these works focused on customer prediction using customer data available with them, which only contains customer stats and has nothing that can refer to the parking capacity and availability. Typical Data set include customer's personal information like name, gender, age, and credit score etc.

In terms of parking prediction there is not a lot work done on the prediction of parking space. In [8] Fabian and Sergio have summarized different methodologies to extract the parking data from different probe vehicles. They collect data from different probe vehicles and apply machine learning to infer the legality and availability of those parking spaces. They use taxi as probe vehicles along with GPS. They use different default sensors installed in taxis to gather parking data. With the help of GPS they map this data onto map. Using this data and map they predict empty parking spaces using Random forest classifier. In [9] Eleni I. Vlahogianni uses the on-streets parking Sensors data to predict the short-term parking availability and short-term occupancy. He has used 5 min data of 1st hour to predict the occupancy for rest of the hour. In [10] Eran and Christopher predicted the price of a parking space using features like its neighbor's parking prices, the past occupancy rate and location features. They then studied its impact on surrounding parking spaces and proposed a price which makes whole surroundings more profitable.

We are working on prediction customer inflow and outflow based on parking data. This is a relatively new Spectrum and not much has been done to predict customer behavior using traffic data.

3 Proposed Methodology

In this paper, we propose an RNN that takes into account traffic flow within a specific traffic parking lot of a departmental store and then predict the traffic pattern inside the parking lot, which will in turn tell store owner customer inflow and outflow for coming days. We propose a RNN for predicting traffic in that specific parking lot based on previous years traffic data.

We encountered a dataset with parking data in format of car parking logs with entry of each car, its exit time and its discount type (which showed if any discount was given to car at parking rates) and its exit time as shown in Fig. 2.

1	Plate No.	In	Out	Discount Type		
2	89더5276	2014-1-2 9:00	2014-1-2 10:33	Normal		
3	66루4088	2014-1-2 9:00	2014-1-2 13:46	Low-Pollution		
4	86두1772	2014-1-2 9:00	2014-1-2 10:19	Normal		
5	52루1172	2014-1-2 9:00	2014-1-2 10:09	Normal		
6	23서4885	2014-1-2 9:01	2014-1-2 9:26	Normal		
7	02므8963	2014-1-2 9:01	2014-1-2 9:19	Normal		
8	64허2177	2014-1-2 9:01	2014-1-2 10:01	Normal		
9	67소3281	2014-1-2 9:02	2014-1-2 14:17	Small		
10	31므2024	2014-1-2 9:02	2014-1-2 9:36	Normal		
11	52서4448	2014-1-2 9:03	2014-1-2 10:24	Normal		
12	33무2232	2014-1-2 9:03	2014-1-2 18:00	Normal		
13	49누5820	2014-1-2 9:04	2014-1-2 9:25	Small		
14	66므8882	2014-1-2 9:04	2014-1-2 9:36	Carfree day		
15	65서6000	2014-1-2 9:04	2014-1-2 10:13	Normal		
16	33두7089	2014-1-2 9:04	2014-1-2 13:30	Small		
17	25우4036	2014-1-2 9:05	2014-1-2 9:26	Normal		
18	12수3902	2014-1-2 9:05	2014-1-2 9:25	Small		
19	09가8141	2014-1-2 9:06	2014-1-2 11:44	Small		
20	경기84나6885	2014-1-2 9:06	2014-1-2 9:43	Normal		
21	경기830아4616	2014-1-2 9:07	2014-1-2 9:07	Normal		
22	15더9780	2014-1-2 9:08	2014-1-2 9:13	Small		
23	경기37거9188	2014-1-2 9:08	2014-1-2 13:13	Normal		
24	37무3556	2014-1-2 9:11	2014-1-2 9:25	Normal		
25	39허3669	2014-1-2 9:12	2014-1-2 13:35	Normal		
26	경기51바2029	2014-1-2 9:12	2014-1-2 9:15	Normal		

Fig. 2. Data format in Source Files (Excel)

The original dataset was provided in multiple files for each year cover variant amount of data. For example for year 2013 data was divided into two different files with first containing 7 months of data starting from 3rd January till 31st 1st August, while second file provided overlapped data starting from 1st of July 2013 till 31st Dec 2013. Interesting fact was that even though the data was extracted from same source system, the format and fields varied among these files as well. For example in the first 7 month file of 2013 dataset, there were an extra field of weekday which was missing from other files. Another anomaly was the date format in each excel file that was different. This data set was provided in form of four excel files. This data was for 2 years spanning 2013 and 2014.

To process data python was our goto language. First these excel files were exported into .CSV format. The reason for this change was to make them easily accessible in python code. Different analysis were performed on this dataset and multiple patterns were observed. The most useful results were produced when the data was aggregated into day based format. So among variant aggregations and combinations we choose day based aggregation. Data displayed in Fig. 3 shows data snapshot in day based aggregation. It was converted into day grain with 6 different features day, incoming cars, outgoing cars, overnight parked cars, weekdays and holiday. We applied feature scaling on these features to remove bais towards any specific feature. In this process we normalized data in such a way that all 6 features were scaled in values between 0 and 1. This process

```
Date, Incoming Cars, OutGoing Cars, Parked Overnight, isWeekday, IsHoliday
1356825600.0,0,0,0,6,0
1356912000.0,36,36,0,0,1
1356998400.0,0,0,0,1,0
1357084800.0,251,251,0,2,1
1357171200.0,251,251,0,3,1
1357257600.0,263,263,0,4,1
1357344000.0,198,198,0,5,1
1357430400.0,0,0,0,6,0
1357516800.0,344,344,0,0,1
1357603200.0,361,361,0,1,1
1357689600.0,374,374,0,2,1
1357776000.0,345,345,0,3,1
1357862400.0,221,221,0,4,1
1357948800.0,129,129,0,5,1
1358035200.0,0,0,0,6,0
1358121600.0,219,219,0,0,1
1358208000.0,247,247,0,1,1
1358294400.0,345,345,0,2,1
1358380800.0,350,350,0,3,1
1358467200.0,349,349,0,4,1
1358553600.0,223,223,0,5,1
1358640000.0,0,0,0,6,0
1358726400.0,389,389,0,0,1
1358812800.0,342,342,0,1,1
1358899200.0,346,346,0,2,1
```

Fig. 3. Data in day based grain (After preprocessing)

ensures that neural network gives equal importance to each feature (initially). If this process is not applied and the scale of different parameters is varying then convergence for neural network becomes problematic. Such was the case in this scenario where some parameters varied in single digits while other were in scale of hundreds. Sample snapshot of normalized data is shown in Fig. 4. This data was then fed to neural network to get predictions for our desired scenario.

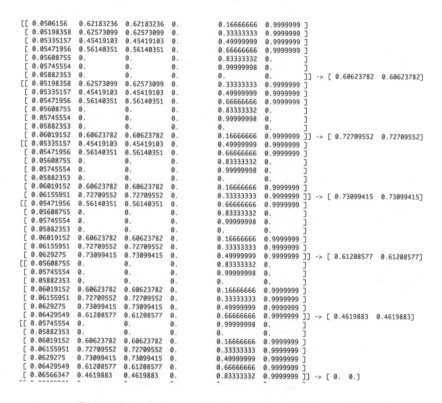

Fig. 4. Data feature scaled for neural network training

We experimented with variant layers of RNN ranging from 3 till 15. The input layer consists of a week's data as an input (7 days into 6 features each) and predicts incoming and outgoing traffic for the next day.

We received the dataset of a parking lot with two years of data. The data was in the form of car logs which contained information about cars and its parking status.

The data was formatted in such a format that each car plate number was given followed by its arrival time. Then next was its departure time from parking and lastly the type of discount it was offered on leaving the parking. For example, we had entries like 12-XX-3456, 10:00:00 AM, 11:55:05 AM, Normal. The first column is the plate number followed by 10:00:00 which is the car's arrival time

in format HH:MM:SS, similar format is followed for third column which is time
when the car left the parking and lastly 'Normal' is the discount type offered to
the car on exit. We received these car logs for 2013 and 2014.

This data was converted from crude car parking logs into useful day-based
information. We extracted six different features from this data. It was arranged
into a format that Date in format YYYY-MM-DD, followed by total incoming
cars on that day, total outgoing cars on parking on that day, overnight cars
parked on that day, day of the week and the lastly weather the day was holiday
or not. Via this summarization we converted over 177000 logs into 729 days.

We then applied feature scaling on all these features to remove bias due to
different value ranges of different features using Eq. 2.

$$NormalizedFeature = \frac{(Feature - MinimumValue)}{(MaximumValue - MinimumValue + e)}. \quad (2)$$

Here 'e' is Euler's constant and it was introduced as an error constant and
to reduce the possibility of division by zero.

We divided this data set into 70:30 for training and testing purpose. The
RNN was trained in 70% or 510 days. For test data we used rest of the 30%
or 219 days. Information related to this dataset is summarized in Table 1 and
system wide architectural overview is shown in Fig. 5.

We experimented on two different variations of RNNs. We used standard
vanilla RNN's and then further used LSTMS to verify and improve our results.

Recurrent neural networks are one subtype of neural networks which have
a memory element attached to them. Their specialty is to capture time series

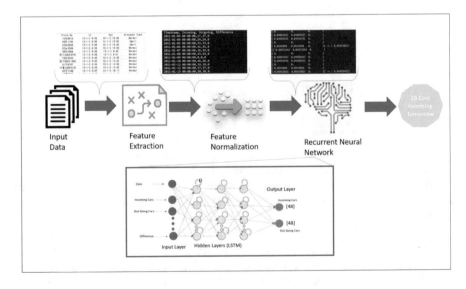

Fig. 5. System architecture overview

Table 1. RNN configurations

Feature	Value
Basic data input type	Car entry logs
Data format	Plate number, incoming time, outgoing time, discount type
Logs start date	1st Jan 2013
Logs end date	31st Dec 2014
Total records	177000 entries
Records span	2 Years (729 Days)
Aggregation level	Days
Train data	70%
Test data	30%
Normalization formula	(Feature − Min)/(Max − Min + e)

dependencies. Our dataset is one example of time series dataset therefore we are using RNN rather than using convolutional neural networks which are very good at classification problems but perform poorly on time series data. We worked with both standard Vanilla and LSTM RNNs. LSTM differ from vanilla RNN's in a way that they have a complicated memory structure which decides for itself that which information it wants to keep and which it wants to discard, as opposed to Vanilla RNN's which have a simple memory.

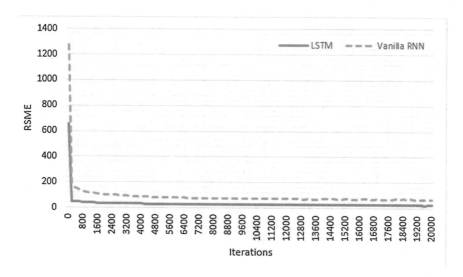

Fig. 6. RSME of Vanilla RNN vs LSTM during training time (optimal values)

Our RNN comprised of an Input layer with 42 input neurons. These neurons were resultant of 6 features of each day with 7 days of data each time. We analyzed multiple configurations with input data per day varying it from 3 to 14. We experimented with various configurations of hidden layers varying from 3 to 14. Output layer contains two neurons predicting incoming and outgoing cars for next day. We also experimented with different learning rates from 0.1 to 0.00001. With different types of RNNs, we had different impact due to learning rate changes as shown in Fig. 6. We also varied different training iterations for the purpose of getting optimal learning and avoiding overfitting at the same time. These properties are summarized in Table 2.

Table 2. RNN configurations

Parameter	Value
RNN type	Vanilla, LSTM
Input layer	1 * [7 days * 6 features] = 42 Neurons
Output layer	1 * 2 Neurons [Incoming, outgoing]
Hidden layers	[3, 5, 7, 10, 12, 14] Layers
Learning rate	0.1, 0.01, 0.001, 0.005, 0.0001, 0.00001
Iterations	1,000, 2,000, 10,000, 20,000

4   Experimental Results

We experimented with different variations and combinations across vanilla as well as LSTM RNN's. Root mean square error or simply RSME was used as error metric. We started with Vanilla RNN's and applied different learning rates. We observed that smaller learning rates were producing better results, less RSME and lower training error. The best learning rate was around 0.00001 and 0.0001. Higher learning Rates tended to overshoot most of the time. RSME also sky rocketed when we used learning rates in higher range like 0.1.

Figure 7 shows the performance of Vanilla and LSTM on testing data in terms of actual traffic on that day compared with the prediction done by vanilla RNN.

Best learning rate was found to be 0.00001 for vanilla RNN's. We wanted to train our RNN as much as possible but also wanted to avoid overfitting. For that we experimented with various iteration cycles. We found that lower training cycles in magnitude of 1000 to 10,000 tend to lead towards undertraining while training cycles of over 20,000 also tend to overfit the data. The best results were found using 20,000 iterations as shown in Fig. 8. For different memory configurations 7 layers were found optimal as there was no improvement using more than seven layers. We were able to reduce RSME to about 0.239902 using vanilla RNN's.

Using LSTM, we also experimented with different configurations. Similarly, Vanilla RNN's higher learning rates tend to overshoot while lower ones were

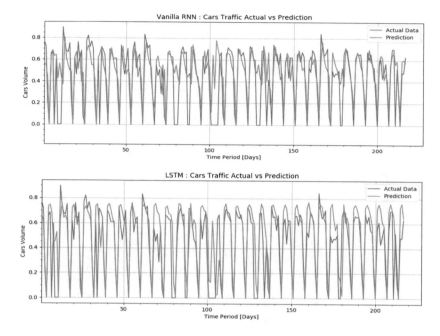

Fig. 7. Actual vs predicted value for next day.

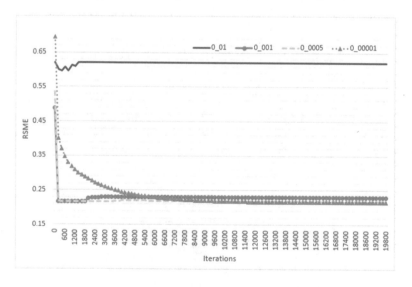

Fig. 8. RSME of Vanilla RNN on different learning rates.

slow to converge. The optimal learning rate was found to be 0.01. The optimal number of iterations were also 20,000 before overfitting started. For the layers we experimented with range of 3 till 15 but most optimal were seven layers for LSTM as well. Interestingly RNN's with LSTM performed much better then vanilla RNN's. We were able to reduce the RSME to 0. 194362 as shown in Fig. 9. This RSME is 0.045 better then Vanilla RNN's which a massive improvement in itself. The reason for this drastic improvement is hidden in the way LSTM works. As LSTMs decide by themselves which information to store and which to discard they can track better information then vanilla RNN's which can only bank on recent memory rather than the whole history.

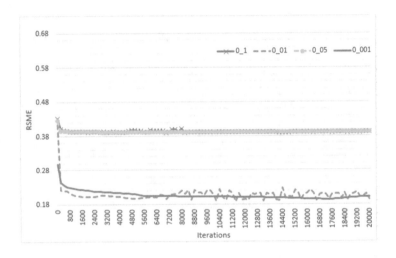

Fig. 9. RSME of LSTM on different learning rates.

We achieved a RSME of 0.194362 which is quite good and is comparable to state of the art results in this regard. The final configuration used to achieve this learning rate was using Learning Rate of 0.01, 20,000 iterations, 7 Hidden Layers and 2 output layers. A comparison of both flavors of RNN is shown in Table 3.

Table 3. Vanilla RNN vs LSTMs.

Configurations	Vanilla RNN	LSTM
Iterations	20,000	20,000
Learning rate (Optimal)	0.00001	0.01
Memory	Standard	LSTM
Optimal number layers	7	7
RSME (Optimal)	0.23993	0. 194362

5   Conclusion

In this paper we proposed a different method of calculating and predicting customer influx and outflux in a departmental store. We proposed an RNN network which based on the data of previous years, predicted parking incoming and outgoing traffic for the next day. Our work used two different variations of RNNs. We concluded that performance of LSTM was better than vanilla RNNs. Our work scenario is particularly useful for large stores like Walmart, Chesco, Emart and Martro. This can help them plan about their resources in advance, based on the predictions and would save resources and cost.

In future work we will dissect this data set even more and try to predict parking incoming and outgoing patterns based on a finer grain. From the perspective of large store owner our work is useful. It can be more helpful if it can predict traffic inflow not only day wise but also on smaller grains like hours. For example, if an owner can know that the afternoon will have low customer inflow but later in the day it will increase many folds. This can help them plan resources more efficiently. There are a lot of environmental factors that impact the traffic and parking trends. These include but are not limited to temperature, humidity and rainfall. We will be targeting these two factors for our future works. Also, we will try to improve the accuracy of these results with some more factors and granular details.

Acknowledgement. This work (*GrantsNo. C*0515862) was supported by Business for Cooperative R&D between Industry, Academy, and Research Institute funded Korea Small and Medium Business Administration in 2017.

References

1. Shoup, D.: Cruising for parking. Transport Policy **13**(6), 479–486 (2006)
2. Caicedo, F.R., Lopez-Pita, A.: Parking management and modeling of car park patron behavior in underground facilities. Transp. Res. Record J. Transp. Res. Board **1956**, 60–67 (2006)
3. Castro-Neto, M., Jeong, Y., Jeong, M., Han, L.: Online-SVR for short-term traffic flow prediction under typical and atypical traffic conditions. Expert Syst. Appl. **36**(3), 6164–6173 (2009)
4. Yu, R., Zheng, S., Anandkumar, A., Yue, Y.: Long-term forecasting using tensor-train RNNs. eprint arXiv:1711.00073, vol. 171100073 (2018)
5. Ishak, S., Alecsandru, C.: Optimizing traffic prediction performance of neural networks under various topological input, and traffic condition settings. J. Transp. Eng. **130**(4), 452–465 (2004)
6. Yu, H., Wu, Z., Wang, S., Wang, Y., Ma, X.: Spatiotemporal recurrent convolutional networks for traffic prediction in transportation networks. Sensors **17**(7), 1501 (2017)
7. Spanoudes, P., Nguyen, T.: Deep learning in customer churn prediction: unsupervised feature learning on abstract company independent feature vectors. eprint arXiv:1703.03869, vol. 03869, no. 170303869 (2018)

8. Bock, F., Di Martino, S., Sester, M.: Data-driven approaches for smart parking. In: Altun, Y. (ed.) Machine Learning and Knowledge Discovery in Databases, ECML PKDD 2017, vol. 10536, pp. 358–362. Springer, Cham (2017)

9. Vlahogianni, E., Kepaptsoglou, K., Tsetsos, V., Karlaftis, M.: A real-time parking prediction system for smart cities. J. Intell. Transp. Syst. **20**(2), 192–204 (2015)

10. Simhon, E., Liao, C., Starobinski, D.: Smart parking pricing: a machine learning approach. In: IEEE Conference on Computer Communications Workshops (INFOCOM WKSHPS), Atlanta, GA, pp. 641–646 (2018)

11. Schmidhuber, J.: Deep learning in neural networks: an overview. Neural Netw. **61**, 85–117 (2015)

12. Pierce, G., Shoup, D.: Getting the prices right. J. Am. Plan. Assoc. **79**(1), 67–81 (2013)

Identifying Topics: Analysis of Crowdfunding Comments in Scam Campaigns

Wafa Shafqat and Yungcheol Byun[✉]

Department of Computer Engineering,
Jeju National University, Jeju, South Korea
wafashafqat92@gmail.com, ycb@jejunu.ac.kr

Abstract. In the Internet era, the effect of reviews and comments is reinforced as one can make a well-informed decision based on the experiences of others. Crowdfunding site, Kick-starter presents a platform to the backers to leave their feedback on a campaign. However, the comments are in abundance and diverse in nature that it becomes barely possible to wade through them to pick up the desired information. This study takes a step to identify the hidden themes in these comments to discover the different topics of discussion in scam campaigns and then these topics are compared with the topics identified in genuine campaigns. Topic models such as LDA (Latent Dirichlet Allocation) have been used in many areas. We have also used LDA to extract the dominant topics in a document. We evaluated this model on both scam and non-scam campaigns comments. We observed that the resulted topics in each category are distant from each other.

Keywords: Crowdfunding · Scams · Topic modeling · LDA

1 Introduction

The burgeoning accessibility and availability of unstructured text data especially the customer's comments and reviews grant public the freedom to express themselves in the most natural way, by not restricting them to follow a specific format or predefined rating scales. They are provided with an opportunity to say whatever they want to say in whatever order or manner they find convenient and appropriate. These reviews or comments are considered to be very effective for other users to make informed decisions. Comments in any nature i.e., positive or negative are effective for other users to help them take a decision. A survey reported astounding stats regarding online reviews, according to them positive online reviews helped 90% of people take decision of buying a product while negative reviews helped 86% of buyers [1].

Many studies have focused on analyzing these reviews e.g., reddit, and twitter, crowdfunding- *a practice of online fund raising*, has emerged recently and there is a need to analyze the comments and backers feedbacks. As the success and popularity of crowdfunding is expediting day by day, maintaining and building investors trust has become a challenge too. The creator of a campaign has obligations like trust and fiduciary towards his backers.

© Springer Nature Switzerland AG 2019
R. Lee (Ed.): SNPD 2018, SCI 790, pp. 137–148, 2019.
https://doi.org/10.1007/978-3-319-98367-7_11

Generally, crowdfunding platforms are comprised of many sections providing a comments section too for backers to leave reviews of product or any comment. For a new backer, these comments can also help in making a decision to invest in that project. These comments, showing actual experience and issues of backers, can reveal many other aspects of a campaign, as sometimes potential issues can be spotted through these comments, which most probably a novice user would never think of inquiring them. However, the comments are in abundance and diverse in nature that it becomes barely possible to wade through them to pick up the desired information. Therefore, straining themes and context out of bulk data available has become a hefty challenge. In order to alleviate this problem, the efforts done in text processing, text mining and analysis have been escalated at bizarre levels, in recent years.

To perceive the meaning of words and enact the context of data is an imperative challenge in text data analysis. Thus, word counts as a brief summary of text, can be perplexing and insufficient for deeper analysis. In this paper, the presented model and analysis is inspired and based on a class of models generally known as "topic" models [2]. These topic models use the content of the comment to acquire the latent set of sentiments or topics hidden inside, each topic is expressed with its own vocabulary.

Topic models provide a simple, yet powerful way to model high-level interaction of words in speech. The meaning of speech arises from the words jointly used in a sentence or paragraph of a document. Meaning can often not be derived from looking at singular words. This is very much evident in consumer reviews where consumers may use the adjective "great" in conjunction with the noun "experience" or "disappointment." When doing so, they may refer to different attributes of a particular product or service.

In this study, we will focus on the comments section of most popular crowdfunding site, kickstarter.com. The projects selected are of two categories, Scam and Non scam. The analysis will be based on the comparison of comments in both categories, by finding the hidden topics of discussion. The key objectives of this work are:

- Understanding scam campaigns based on backers reactions
- Finding ways to automatically classify scam projects
- Finding underlying topics of discussion among backers

The rest of the paper comprises of literature review in Sect. 2, targeted research questions and dataset curating process is explained in Sect. 3. In Sect. 4, LDA process is explained precisely. Section 5 being the primary section presents the analysis of comments through the complete process of LDA and closes with the identified types of comments. This work is concluded in Sect. 6.

2   Research Questions and Dataset

Our study focuses to find out what role can comments of backers play in understanding the creator's attitude towards the campaign. A screen shot of a sample kickstarter campaign page along its comments section is shown in Figs. 1 and 2 respectively. We want to focus on the comments section in order to have some clues regarding the concerns of backers. The key focus is on following questions:

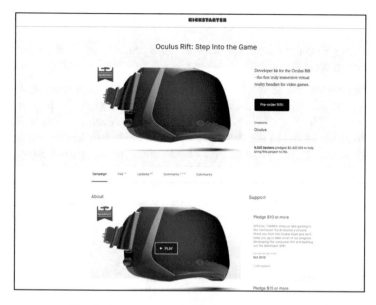

Fig. 1. A screenshot of kickstarter campaign page

Fig. 2. Comments section of a kickstarter campaign

RQ1. What types of comments backer leave after the expected product delivery date?

RQ2. Comparison of themes identified in scam projects with backers comments on non-scam projects?

For data collection, as we have a list of projects for both scam and non-scam categories. This list of projects comprises of project title and IDs. All project IDs are of kickstarter.com. The IDs of listed projects were first collected and then for each of these projects we extracted the comments. We were able to collect comments of 200 projects in total, i.e. 100 projects of each category. These projects were randomly selected. Project creator can also respond to backers queries in the comments section. We initially collected all the backers' comments and creator comments as well. For the purpose of this research we focused only on backers comments. Each comment is collected along with the backer's name, time when the comment was made, and the comment content.

3   Related Work

In recent years, there has been lot of research work done on crowdfunding, predicting success and failure [3], dynamics of crowdfunding [4], and content analysis [5] etc. Due to the substantial challenges of legitimacy and fraudulent activities observed by these platforms, people find themselves at risk while pledging in a campaign. To study different dynamics of scam campaigns, one way is to analyze how people react through their comments. These comments are being analyzed through discovering different topics out of these comments and then comparing them with legit campaigns.

In order to extract different themes and topics out of large text data sets, topic models such as LDA (Latent Dirichlet Allocation) [2] have been used in many areas. In [6], online reviews are being focused and multi grain topics are being extracted by applying extensions to basic LDA model. There have been other extensions of topic models as labeled, where one or multiple labels are associated with a single document [7], partly labeled [8], constrained [9] etc., models. In [10], researchers have identified hidden topics in project updates periodically posted on crowdfunding campaign's update section.

Online Topic Model (OLDA), is being presented in [11], a topic model for identification of thematic patterns and developing topics in text and also captures the change in themes and topics over time. The proposed approach in this paper enables the topic modeling schemes to perform in an online manner, i.e. LDA to additively generate an updated model. Twitter is a source of abundant text data; different trends and tweeting patterns can be observed based on different events. In [12], a joint Bayesian model is introduced that focuses on the following two intrinsic research problems in an inter dependent manner; one being the fact that large data sets can be more helpful for analysis and research if topics are being extracted from text tweeted on a particular event and second is to segment the events. The proposed model in a single unit performs topic modeling and event segmentation.

During the past decade, training to label web and text documents has comprehensively attracted many researchers. For a lot of classification problems, many learning methods e.g., Naïve-Bayes, support vector machines (SVM), and k nearest neighbors (k-NN) have attained decent results [13, 14]. Zhao [15], has tried to separate the topics from sentiment words through a switch. Besides that, we can also find work aiming at modeling debate discussions [16] and figuring out attitudes or stances in debates [17]. In [18], the relationship between blog post and comments on it has been modeled. Classification and modeling or detection of bogus reviews and reviewers has been done in [19]. In [20], Gretarsson has presented TopicNets based on statistical topic models, for the analysis of large text corpus. There have been many different kinds of visualizations and techniques like iterative topic modeling, visual filtering etc. are introduced in order to back knowledge discovery. The following objectives have been targeted in [21]: use of topic modeling to discover important and applicable public health topics, discovering related themes to obesity, identifying spatial pattern in themes, etc. The target was achieved by using LDA for topic modeling and Geographic Information System (GIS) was used for spatial analysis.

4   Latent Dirichlet Allocation (LDA)

Latent Dirichlet allocation (LDA) is a generative probabilistic model of a corpus [2]. The primary purpose of LDA is to represent the documents as random mixture over latent topics, and these topics are indicated by a distribution over words. The basic idea is that documents are represented as random mixtures over latent topics, where each topic is characterized by a distribution over words. Figure 2 shows the generative process of LDA.

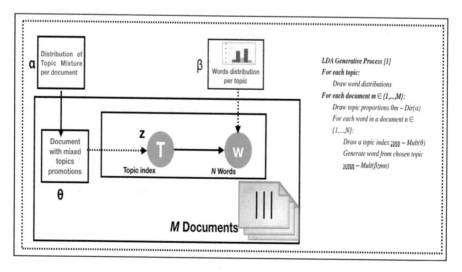

Fig. 3. Generative LDA process

In Fig. 3, we can see that for each document we have an association with each topic by some probability. There are two things to observe; first, for each document we have some rated probabilities of topics. Similarly, every topic has some rated probability of words. Therefore, there are possibilities that each document might have a footprint of every topic. However, we are only concerned with the top ones, usually top three or four topics.

5   Analyzing Comments

In order to determine the themes or topics categories in Kickstarter comment section, Latent Dirichlet Allocation (LDA) was applied on the dataset, LDA is an unsupervised generative technique that is generally applied to identify the latent topics along with associated words with that particular theme in a specific document. Through LDA, massive amounts of unlabeled documents are being analyzed by clustering the frequently co-occurring words. Following are the steps performed.

5.1   Comments Selection

We were faced with some challenges before applying the LDA. In order to overcome these challenges we had to find the answers to the following critical questions:

(a) Is it a good Idea to use all the comments of a project for themes extraction?
(b) For projects of two categories (Scam/Non Scam), what will be an effective criteria for comments selection?
(c) How many comments should we select per project?

In order to answer these questions, we made the following criteria:

- To check the legitimacy of a project, the expected delivery date is an important factor. As, if a product is not being delivered so far even its delivery date has passed long ago, we can't be sure if this project is legit or not. Therefore, we selected comments after the expected delivery date of a project.
- Top 50 recent comments were selected for each project, as number of comments were not evenly distributed among all projects. Some projects had many comments while other had a few. All the comments were selected for the projects having fewer than fifty comments.

5.2   Data Cleaning

Before applying LDA, all URLS and stop words were filtered out. All the comments from the creator of the project were also removed. This step is also known as data preprocessing.

5.3   Applying LDA

In order to apply LDA, we used each comment as an individual document. Each document generated a pool of topics from which we selected top relevant topics, see

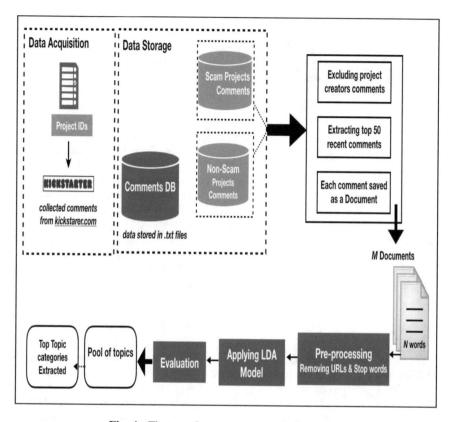

Fig. 4. The complete comments analysis process

(Figs. 3, 4). In Fig. 4, the complete process of our system is described i.e., from the data acquisition stage leading to data storage, then preprocessing, model implementation and eventually the themes or topic extraction stage. Data for scam and non-scam project comments is stored separately. The comments were considered as a document for the LDA model. All the data or more precisely all documents comprising of individual comments are stored in text files.

5.4   Types of Comments

The themes or topics identified in the comments benefit us to have a bird eye view comparison of backer's discussions in scam and non-scam campaigns. The primary purpose of using these comments is to analyze the backers views regarding a project. As, the comments and reviews of backers can help us identify different aspects and behaviors of backers towards a specific product and it can help other backers to take a well aware decision. In Tables 1 and 2, we can see the topics categories identified in scam campaigns and non-scam campaigns, respectively. These tables present the topic categories, along with the dictionary words, and examples for each category.

Table 1. Discussion topics in comments section of scam campaigns

Topic category	Words and examples	
	Dictionary words	Examples
Waiting for rewards	Fulfil, rewards, waiting	• "I invoke my rights under Kickstarter's Terms of Use, Project Creators are required to fulfill all rewards of their successful fundraising campaigns or refund any Backer whose reward they do not or cannot fulfill. I demand a full refund for my pledge amount." • "I doubt it, but any new information on when we might be expecting our rewards?"
Asking for refunds	Fulfil, refunds, money, creator, project	• "I HAVE ASKED REPEATEDLY AND RECEIVED NO REPLY. I HAVE POSED IT OPENLY AND IN PRIVATE MESSAGES. I sent a SASE. When should I expect my refund? Thanks." • "I sent a SASE. When should I expect my refund? Thanks."
Waiting for an update/Reward	Waiting, money, refund, update, doesn't, received	• "You can issue me a refund. Thank you very much. I won't email you any information to your Gmail account right now because I don't trust you with any information about me. When the refund is ready, I will rent a PO Box so that you have a place to mail a money order. Once I have my money back, I'll consider about forgiving you personally, as you have asked me to do."
Reporting or taking legit actions against it	Attorney, Kickstarter, project, report, actions response, state, legal	• "No problem, Gary! It might also be worth a minute to fill out a complaint with your Attorney General: As we build up momentum, hopefully some state's attorney general will take Washington's lead and file a consumer protection action/lawsuit against Tom Baker for abusing Kickstarter." • "Hi Phillip, How did you go about reporting this to the Attorney General? I am not sure where to begin."

(continued)

Table 1. (*continued*)

Topic category	Words and examples	
	Dictionary words	Examples
Product never received	Never, still, product, received, mine	• "I donated almost two years ago and never received what was promised." • "Still no magnets." • "Still nothing…" • "Nothing for me as well. Feeling scammed."
Showing Anger ….. or Disappointment	Fraudster, what, why, legally..	• "What ever happened to all the boxes of manufactured air track without the trolleys that was in update #7 June 2013, why haven't these been shipped? I can make up my own bloody trolley if you send me the wheels and bearings. The main thing that I could use is that and that was the unique part of this campaign, not the trolley. Just send me that working part and refund half my money, keep in mind that I have now finished two separate contacts and now working in my third country waiting for this entire time!!" • "If you have not received a refund, you won't. The statute of limitations is 3 years and has expired. The fraudster is legally off the hook. I only do kickstarters from people or companies I know… And even then there can be problems."

In Table 1, only the main topics are listed:

(1) The backers are waiting for the promised rewards and are concerned when those rewards will be received.

(2) In some comments backers are asking for their refunds which is a reflection of their disappointment.

(3) Some comments reflect the anxiety caused due to protracted silence from creator's side resulting in asking for an update or rewards.

(4) In some cases, when backer's frustration level touches the peak, they start taking legal actions against the creator.

(5) It can easily be observed in comments that backers are complaining about that they haven't received the item since yet.

Table 2. Discussion topics in comments section of non-scam campaigns

Topic category	Words and examples	
	Dictionary words	Examples
Product shipment	Product, shipped, wedge, idea	• "Are we supposed to receive confirmation of shipment?" • "Has all the rewards shipped out? Still didn't get a shipping confirmation email."
Product description	Brewer, cup, drink, work, lid, router, device	• "I, too, am interested in a new lid. Would have been nice if it was given to/offered at a discount to backers, since the original lid was poor. Nonetheless, I am quite content with my brewer; while I never drink from it, I do use it to make my coffee :)"
Product's working status	Apps, device, great, support, ads	• "I've had some requests for the firmware, so I've uploaded it to a google drive account. It should be able to sit there pretty much forever if you need it."
Product received	Cards, today, mine, received, loved	• "À perfectly wrapped package with 6 decks arrived in Belgium today, with an additional € 24,- customs ticket. But the decks make up for it. Gorgeous quality. Thanks for this fantastic project." • "I got my unit today!"
Showing excitement	Pledge, received, mine, cards, deck, decks, loving, loved, great	• "Love my deck of cards. One of my favs!" • "Still waiting for my package, cant wait :D"

(6) One of the biggest motivations for a backer to invest in a campaign, is to get that product or rewards, but not getting it till the expected date, causes anger and disappointment in them.

Unlike in Table 1, we can see different categories listed in Table 2. In non-scam campaigns comments, we found that mostly comments are related to the product, following are the top discussion topics found:

(1) The backers are inquiring about the product's shipment status.
(2) As non-scam campaigns are those for which we have a verdict that these have been delivered to the backers as per promise, therefore, backers are mostly talking about the product, how it looks or works etc.
(3) Also the working status of the product is being discussed like if its working well, or if it has some issues.

(4) It's not surprising to find comments confirming that they have received their product or rewards.

(5) As, backers are using the product, they are also exhibiting their emotions and excitement towards the product.

6 Conclusion

Crowdfunding being popular and successful in recent years, comes along many challenges as well. Identifying the legitimacy of a project is equally important and challenging in order to save backers from being scammed or conned. In this study we tried to do a comparison of comments being posted on scam and non-scam campaigns. In natural language processing, themes and topic extraction form text corpus is very common. There are methods like LDA to find hidden topics in raw text data. We have used the comments for both the scam and non-scam comments to identify the hidden topics in them. We observed a clear difference between the comments topic categories. Projects in scam category, received comments mostly portraying the negative emotions, anxiety, and frustration, which in some cases has led them to take legal actions against the campaign owner, and in some cases make them force the backer to return their money or deliver the rewards. On the other hand, in comments of non-scam campaigns, to our expectations reflected that backers are discussing the product itself, e.g., if they liked it or not, etc., confirming that they have received the product.

In future, we aim to extend this work to do this analysis at a deeper level and using more data. This is a first step towards comments analysis, many other factors are to be included for in depth and comprehensive analysis.

Acknowledgement. This work (Grants No. C0515862) was supported by Business for Cooperative R&D between Industry, Academy, and Research Institute funded Korea Small and Medium Business Administration in 2017.

References

1. https://marketingland.com/survey-customers-more-frustrated-by-how-long-it-takes-to-resolve-a-customer-service-issue-than-the-resolution-38756
2. Blei, D.M., Ng, A.Y., Jordan, M.I.: Latent Dirichlet allocation. J. Mach. Learn. Res. **3**, 993–1022 (2003)
3. Cordova, A., Dolci, J., Gianfrate, G.: The determinants of crowdfunding success: evidence from technology projects. Procedia-Soc. Behav. Sci. **181**, 115–124 (2015)
4. Mollick, E.: The dynamics of crowdfunding: an exploratory study. J. Bus. Ventur. **29**(1), 1–16 (2014)
5. Marom, D., Sade, O.: Are the life and death of an early stage venture indeed in the power of the tongue? Lessons from online crowdfunding pitches (2013)
6. Titov, I., McDonald, R.: Modeling online reviews with multi-grain topic models. In: Proceedings of the 17th International Conference on World Wide Web, pp. 111–120. ACM, April 2008

7. Ramage, D., et al.: Labeled LDA: a supervised topic model for credit attribution in multi-labeled corpora. In: Proceedings of the 2009 Conference on Empirical Methods in Natural Language Processing: Volume 1-Volume 1. Association for Computational Linguistics (2009)

8. Ramage, D., Manning, C.D., Dumais, S.: Partially labeled topic models for interpretable text mining. In: Proceedings of the 17th ACM SIGKDD International Conference on Knowledge Discovery and Data Mining. ACM (2011)

9. Andrzejewski, D., Zhu, X.: Latent Dirichlet allocation with topic-in-set knowledge. In: Proceedings of the NAACL HLT 2009 Workshop on Semi-Supervised Learning for Natural Language Processing. Association for Computational Linguistics (2009)

10. Xu, A., et al.: Show me the money!: an analysis of project updates during crowdfunding campaigns. In: Proceedings of the SIGCHI Conference on Human Factors in Computing Systems. ACM (2014)

11. AlSumait, L., Barbará, D., Domeniconi, C.: On-line LDA: adaptive topic models for mining text streams with applications to topic detection and tracking. In: Eighth IEEE International Conference on Data Mining, 2008, ICDM 2008. IEEE (2008)

12. Hu, Y., et al.: ET-LDA: joint topic modeling for aligning events and their twitter feedback. In: AAAI, vol. 12 (2012)

13. Baldi, P., Frasconi, P., Smyth, P.: Index. Wiley, New York (2003)

14. Sebastiani, F.: Machine learning in automated text categorization. ACM Comput. Surv. (CSUR) **34**(1), 1–47 (2002)

15. Zhao, W.X., et al.: Jointly modeling aspects and opinions with a MaxEnt-LDA hybrid. In: Proceedings of the 2010 Conference on Empirical Methods in Natural Language Processing. Association for Computational Linguistics (2010)

16. Mukherjee, A., Liu, B.:. Mining contentions from discussions and debates. In: Proceedings of the 18th ACM SIGKDD International Conference on Knowledge Discovery and Data Mining. ACM (2012)

17. Somasundaran, S., Wiebe, J.: Recognizing stances in ideological on-line debates. In: Proceedings of the NAACL HLT 2010 Workshop on Computational Approaches to Analysis and Generation of Emotion in Text. Association for Computational Linguistics (2010)

18. Hasan, K.S., Ng, V.: Stance classification of ideological debates: data, models, features, and constraints. In: Proceedings of the Sixth International Joint Conference on Natural Language Processing (2013)

19. Mukherjee, A., Liu, B., Glance, N.: Spotting fake reviewer groups in consumer reviews. In: Proceedings of the 21st International Conference on World Wide Web. ACM (2012)

20. Gretarsson, B., et al.: TopicNets: visual analysis of large text corpora with topic modeling. ACM Trans. Intell. Syst. Technol. (TIST) **3**(2), 23 (2012)

21. Ghosh, D., Guha, R.: What are we 'tweeting' about obesity? Mapping tweets with topic modeling and Geographic Information System. Cartogr. Geogr. Inf. Sci. **40**(2), 90–102 (2013)

Security Monitoring Technological Approach for Spear Phishing Detection

HooKi Lee, HyunHo Jang, SungHwa Han, and GwangYong Gim[✉]

Department of IT Policy and Management,
Soongsil University, Seoul, South Korea
hk0038@korea.kr, jangh2@daum.net, taifanz@naver.com,
gygim@ssu.ac.kr

Abstract. Recently, security accidents due to intelligent attacks such as (APT, Advanced Persistent Threat) attacks have been widely occurring in various industries. Spear-phishing using e-mail is mainly used because it is necessary to enter an internal system or a network in order for such a targeted APT attack to be successful. Most organizations build and operate e-mail exclusive use security systems, but since they rely on dynamic analysis of attachments or blacklist-based detection, normal link-style spear phishing attacks in various texts may be bypassed. Therefore, as we cannot say that the existing security control system based on e-mail security solution is effective, this study is going to examine the problem current e-mail security solution and present an improved proposal model based on security control.

Keywords: Security monitoring · Spear phishing · Hacking mail
Phishing mail · Spammail

1 Purpose of the Research

As countries, institutions, corporations and individuals become more reliant on ICTs throughout the society, such as storing information and knowledge on various types of computers and providing services, the malicious behaviors to seize, tamper and destroy integrated information and knowledge, and its method is also becoming intelligent. As information security technology evolves, the attacker specifies clear targets to raise the performance and probability of success of cyber attacks, and intelligently attack spending much more preparation and times, and this type is called intelligent persistent threat (APT, Advanced Persistent Threat) attack. In order for an APT attack to be successful, it must enter an internal system or network. However, as organizations having a lot of hard work and an integrated information or information infrastructure worth to implement APT attacks are well implemented with security measures, an attacker circuitously attack users having access authority to confidential data and important system, such as major directors and employees, instead of direct attack on the system or network. An attack method mainly used at this time is spear-phishing using e-mail [1]. 91% of targeted attacks begin with spear phishing e-mails, 94% of spear phishing e-mails attach files, and 76% of attack targets were found to be corporations or

© Springer Nature Switzerland AG 2019
R. Lee (Ed.): SNPD 2018, SCI 790, pp. 149–163, 2019.
https://doi.org/10.1007/978-3-319-98367-7_12

government agencies [2]. Spear phishing e-mails contain malicious files or URL links, and what's more important, they misrepresent an acquaintance by using social engineering techniques or deceit the user with content that is relevant to the user's work.

According to one survey, more than 50% of spear phishing victims were clicking on attached files. Also spear phishing attack costs 10 times of general phishing or spam, and the success rate is 10 times the general phishing or spam [3]. Once an attacker succeeds in initial penetration, it can be seen also in the case of APT attacks such as 3.20 cyber terror, Korea Hydro & Nuclear Power, and Interpark that there is a limit in responding quickly and effectively in standpoint of defenders. In the end, early detection is more important than post-response, and many organizations that perceive spear phishing as the most serious threat are building and running a defense system with e-mail-related security solutions. Today's security solutions provide various functions such as blocking spam, analyzing/blocking malicious behavior of attachments, blocking black list-based on sender IP/account or harmful IP/URL, but, in fact, as the security solution cannot properly detect spear fishing e-mails, defense system built solely with the solution it cannot be said that security system built with solution is effective from a security management point of view.

So this study is going to propose an effective spear phishing system based on security control by analyzing the current status and problems of hacking email including spear phishing, and verifying with applying of actual cases.

2   Current Status and Problems

2.1   Hacking Mail Distribution Status

According to IBM's X-Force researchers, half of all e-mails are estimated to be spam, and hacking e-mails with malicious code or URLs are increasing rapidly among all spam [4] (Fig. 1).

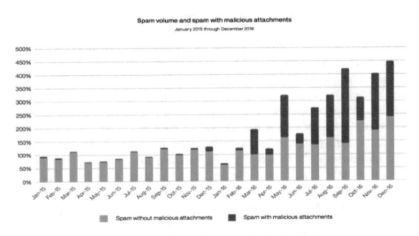

Fig. 1. The proportion of spam with malicious files [4]

According to the 2018 Data Breach Investigations Report(DBIR) released by Verizon Enterprise Solutions, analysis of informations from 67 organizations and 53,000 incidents and 2,200-odd breaches in 65 countries revealed that Phishing were used in 90% of cyber attacks. Google and International Computer Science Institute selected Phishing, password reuse and key logging as the top three cyber threats, and they emphasized Phishing posing as trusted people or organizations to approach goals and to steal information was the most threatening [5].

Therefore, hacking mail with malicious code or URL attached, and spear phishing thoroughly personalized by applying social engineering technique is a universal, easy, and highly successful penetration means for attacker to perform APT attack.

2.2   Characteristics of Hacking Mail

Evolved hacking mails like spear phishing have several features as follows: Before sending spear phishing, the hacker collects information about the attack target. Along with information on the target organization, it collects personal information of users (employees, partner employees, etc.) who have major authority after searching the black market, SNS (Social Network Service) site, homepage, publications, portal (Google etc.), employment site, and so on. It also obtain e-mail addresses of the target for attack mostly via the web, or combine the name (or English initials) with the mail domain of such target person [6].

When the information is accumulated, the attacker creates and sends a hacking mail optimized with, such as, the sender account, title, text composition (including URL link), and attachment file (type and file name) for the target person. When an attacker launches an attachment or URL link of the mail and installs malicious code, the attacker grasps the main system and network of the target organization through this.

The attacker mainly uses malicious code that exploits the zero-day vulnerability to avoid being detected by the security solution built in the target organization. Most of the attackers do not directly look for vulnerabilities in systems and applications, but they wait for security researchers or software vendors to disclose vulnerability information and use by producing malicious code using the vulnerability before a company or organization responds [7].

It takes an average of 11.6 days for a top-notch anti-virus vendor to create and distribute signature for detection of new malware, and it is enough time for an attacker to create and distribute malicious code using the new vulnerability [8].

The main types of files attached to hacking mail were examined to be .rtf, .xls, .zip, .pdf, and so on. .exe, a common executable file type, is rarely used because it is highly probable to be detected by existing security solutions, but it is sparsely used in compressed form (.lzh, .rar, .zip). It bypasses the security solution by encoding password on these compressed files [5]. However, in the case of an executable file, it is mostly written in a file of a general document type (.xls, .pdf, .doc, .hwp, etc.) since it may be suspicious or noticed by the user. In this case, it is a highly likely that the user will not be able to recognize the document file even if it is opened, and it is difficult to detect malicious code executed like this because it exploits a zero-day vulnerability. The URL of the hacking message text is linked to a phishing site (or malicious code circulation site), and about 296,000 phishing sites have been found around the world by the third quarter of 2017, and the number is increasing rapidly [9].

2.3 Current Status and Problems of Hacking Mail Response

An important technical method for hacking e-mail response is the function to detect and block attached malicious code, malicious URL/IP, attacker account or IP. Most solutions include the function to detect malicious code using zero-day vulnerabilities hidden in document files such as MS Word, HWP, and PDF, and they are primarily detecting malicious code based on behavior in a virtual machine environment, and hacking mails known through blacklist-based signatures. Table 1 summarizes the key features of the major e-mail security solutions, and most solutions include malicious code detection using zero-day vulnerabilities and are primarily detecting malicious code based on behavior in a virtual machine environment.

Table 1. Key e-mail security solutions [10]

Company	Main function
Ahnlab	• Analyze new malicious code based on behavior analysis and content analysis in virtual machine environment • Detect non-PE malware code such as document files • Minimize error detection by pre-classifying known normal files • Analyze Network-based abnormal traffic and extracted files
Trend Micro	• Detect known or unknown document vulnerability attacks embedded within PDF, MS Office, HWP and other document formats • Detect additional malware downloads but in-depth investigation and analysis through the sandbox • Detect Zombie PC detection, malicious code in advance which enters into malicious URL blacklist-based internal network
FireEye	• Detect malicious code using vulnerabilities of unknown OS, browser and application program based on virtual execution engine • Detect malicious code hidden in common files and multimedia content • Block vulnerabilities and C&C server connections that exploit buffer overflows, provides callback communication information used for data leakage • Prevent malicious URLs in real-time and identify other users within organization that became the attack target
Websense	• Function to prevent intelligent type malware threat such as zero-day attacks • Block data leakage through web traffic • Function of Sandboxing-based malware analysis and forensic reporting
Symantec	• Grant Provides safety ratings for all executables in real-time as well as malicious files through file reputation algorithm when downloading and scanning files • Detect signature and heuristic based malware code when files are stored on your PC • Detect and block through behavior-based analysis when a file is executed

Most hacking mail detection defense solutions isolate attachments and EML and perform sandboxing based dynamic analysis on the files. It performs various analyzes such as network traffic analysis, system call, file change, and analysis of non-compliant content, and some malicious codes are intelligently written to prevent the detection of a virtual machine environment. However, in addition to these dynamic analyzes, countermeasures against phishing sites linked to URLs consist mainly of defensive forms (blacklist addition) after attacks occur. These detection methods are based on the life cycle of phishing sites and the degree of update Which may degrade performance or stability [11]. In addition, in the virtual machine environment, the function of analyzing the URL with the dynamic engine has many false positives and many cases where effective detection is not performed. Figure 2 is a hacking e-mail hacking on mass mailings on March 22, 2017. The link itself is normal behavior in the dynamic analysis of virtual machines, and three domestic and foreign security vendors (companies A, C and S) Blacklist DB of the user. Figure 3 is a hacking e-mail that was circulated on

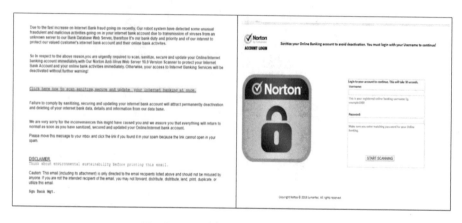

Fig. 2. Phishing mail body and links

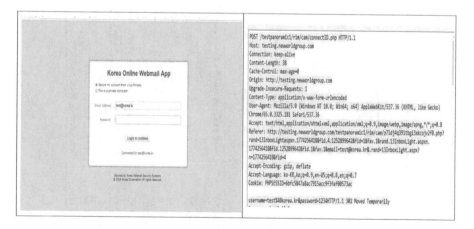

Fig. 3. Phishing site and information

25

Table 2. Key e-mail security solutions [10]

Date of occurrence	Mail title	Number	Outbound IP	Malicious link URL
April 15	Re-activationRequiredfor	381	192.254.234. ***	Http://ba***owto.com/update/Webmail/index.********
April 24	Have you logged into your Kakao account with a new device or IP address?	380	195.154.46. **	Https://cr2***entos.com.br/kak*****
March 23	Warning someone tried logging into your e-mail	175	173.254.28. ***	Http://ho****autolink/autolinkauto/mailboxx/mail****
April 10	Mailbox Upgrade Required	165	52.255.39. ***	Http: //204.152.219.**: ** amail-server /login.php
April 09	Account Verification Notification [important]	109	192.185.197. ***	Http: //arizona.***p-includes/network/thumb/inde*****
March 09	Security Update	83	52.15.194. **	Https://459****ok.io/nid.naver.com/login.php? c****
March 19	Your mailbox will be shut down after 24 h	54	67.227.142. **	Https: //www.cacc****m/wp-includes/theme/n2/*****
March 22	Important! Request confirmation	40	162.144.12. **	Http: //testing.newworld***stpanoramic1/ri****
March 20	Verify [Mail Account] MailBoxAccount	36	192.130.146. ***	Https://lahim****/wp- (https: //sarcasm **** / wp-)
April 19	[Mail account] 暫 stop	34	178.159.36. ***	Https://***.gq/cn/logic/?Email=
March 15	Your mailbox quota is almost full	26	185.204.216. ***	Http://provis*****update-accountverificat***
April 30	FINAL_WARNING_YOUR ACCOUNT WILL BE CLOSED	23	192.185.30. ***	Http://abacus.***nmmm/webupdate/index.php
April 03	SECURITY ALERT	20	192.185.150. **	Http://iodic***.com/now/net/ge***
March 08	Email Security Team	20	185.5.136. **	Http://lea****.net/qqqq/signin/
March 11	Interested in your products.	18	122.224.226. ***	Http://www. ***ree.com/wp-content/*****

April 24, 2017 and contains a phishing site where personal information is leaked by abusing the link inserted in the text, and malicious code is spread through multi-way stoppage, Vaccines, and security vendors' blacklists.

From the end of January to the end of April, 2017, the research environment that operated the email system for about 4 months, the RBL-based spam blocking system, the sandbox-based dynamic analysis, and the three blacks (A, C, S) As a result of building and operating a list DB environment, 952 kinds of spear phishing e-mail using links in the text were diverted to the mail account users by bypassing the security equipment that was installed. Most were identified by e-mail circulating for the purpose of malicious code infiltration of personal information leak, route stop and euphoria. Table 2 is the top 20 information among the number of infiltrated mail, and the damage which is distributed to all users was confirmed after 1–2 days after the user's report.

According to the virus bulletin which provides the virus detection rate information such as the vaccine history, it is confirmed that the link type of the text is normal but the information is not updated in the blacklist DB, Are overwhelming the blacklist signature update of the vaccine engine.

In conclusion, distribution of phishing, malicious code using a link was a clear false negative error, as it has a serious problem of most of them being delivered to users, unless the blacklist DB of the security solutions such as e-mail RBL and vaccine is updated. In the case of attached and link files, various packers of hackers are used to constantly modify malicious files and produce variants to minimize the rate of signature detection of security products, which is efficient for analysis based on email-only dynamic analysis. However, a method of using link is a post-passive method that can be verified only after blacklist DB is updated through relevant sample analysis after the related damage is detected and reported, that is, after damage. Unlike false positive errors, where the results of the detection can be reviewed by specialists such as security guards, false negative errors have a serious risk that the malicious activity itself can be concealed. Therefore, this study propose a model to collect and monitor links of e-mail based on security control in addition to security solution monitoring.

3 Research Model

3.1 E-mail Security Control Model

In Sect. 2, this study could confirm that spear phishing sent by a hacker has limitations in detecting attached files and blacklist matching. Therefore, this study presents the e-mail security control model as follows to detect spear phishing through links or attached file of incoming e-mail text.

This research model largely performs two functions in order to respond positively to spear phishing. First, it is a known function that performs signature-based static analysis and dynamic analysis of headers, mail text, and attached file. Second, it consists of a link extractor for detecting spear phishing, which is disguised as a normal link emphasized in this study.

The incoming e-mail from outside is first parsed by the parser and separated into headers, mail text, and attached file. As the header contains information about the hacking mail attacker, such as sender address, transmission IP, and routing IP, and stores it in the EML Analysis DataBase. The information collected like this is compared with the attacker information of the known hacking mail accidents, and additional analysis is performed when they match. Next, whether or not an external link is included in the body of the mail is checked, and it is separately extracted and stored when it is included. In the case of targeted spear phishing attacks, it leads access to a finely crafted phishing site is made through a link inserted in the body of the message to bypass the detection of automated security equipment, and collection of these link information is essential against the case of security equipment being not included in blacklist DB (Fig. 4).

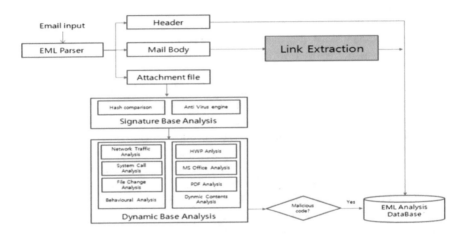

Fig. 4. E-mail security control model

Dynamic analysis of attached file is performed by pattern-based analysis, which is a method of comparing the hash value of a known malicious code and a unique character string in a file to determine whether the file is malicious. The new and variant malicious code, which has not been confirmed here, conducts the behavior analysis through the virtual machine in the next step, dynamic base analysis (Fig. 5).

Analysis of main dynamic network traffic performs monitoring and analysis of operation status on network, such as communication with command control server and downloading additional malicious file when executing target file (Fig. 6).

It records the APIs and memory information that are called when the file is executed through the system call analysis, and records the history of creation, change, and modification of the file through file change analysis. In the case of non-execution type documents, dynamic content analysis is used to detect maliciousness of specific document files that exploit vulnerabilities in certain application programs such as HWP, DOC, and PDF (Fig. 7).

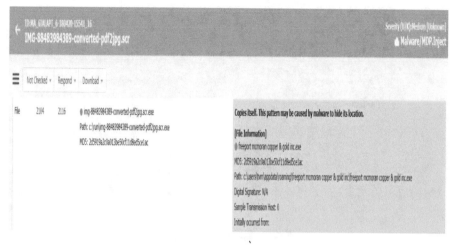

Fig. 5. File change analysis

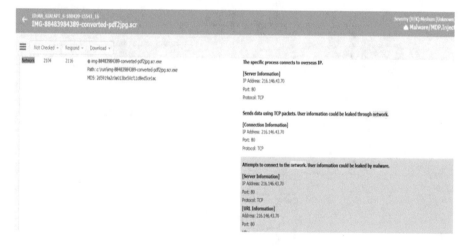

Fig. 6. Network traffic analysis

3.2   Presenting a Link Extractor Model

The link extractor, which is the core of this research model, is a structure for detecting malicious links that are often considered to be normal by security equipment, and it stores URL, image/file links after extracting from the mail body separated by the parser as shown in Fig. 8 after removing de-duplication. It is a basic model that can store all the links of incoming emails and monitor the security through black list and whitelist basis. Link extraction module program that extracts the corresponding action Fig. 10 and DB storage module are composed as in Fig. 11.

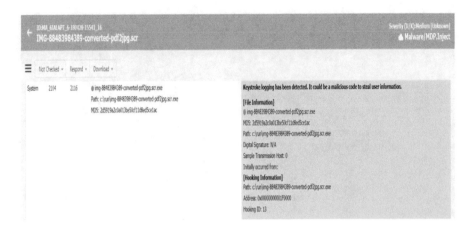

Fig. 7. System call analysis

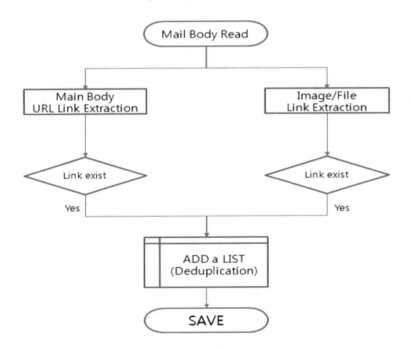

Fig. 8. Structure of link extractor

Import java.sql.Connection	import java.util.regex.Matcher
Import java.util.regex.Pattern	import org.apache.log4j.Logger;
Import org.apache.log4j.PropertyConfigurator;	
Import java.util. *;	
Import java.io. *;	import javax.mail. *;
Import javax.mail.internet. *;	log4j-1.2.15.jar
Commons-pool2-2.3.jar	mariadb-java-client-1.1.9.jar
Ojdbc14.jar	commons-dbcp2-2.2.0.jar

Fig. 9. Status of used library

```
public static void urlExtraction(String strID, File emlFile) throws Exception{

            Properties props = System.getProperties();
            props.put("mail.host", "smtp.dummydomain.com");
            props.put("mail.transport.protocol", "smtp");
            Session mailSession = Session.getDefaultInstance(props, null);
            InputStream source = new FileInputStream(emlFile);
            MimeMessage message = new MimeMessage(mailSession, source);

            String strMessage = new String();
            String strRdate   = null;
            Object content = message.getContent();
            try
            {
                    if (content instanceof String)
                    {
                            strMessage = (String)content;
                    }
                    else if (content instanceof Multipart)
                    {
                        Multipart mp = (Multipart)content;
                        BodyPart bp = mp.getBodyPart(0);
                        strMessage = (String) bp.getContent();
                    }
            }catch(Exception e)
            {

            }finally{

            }
            System.out.println("strMessage " + strMessage);
            String strLink = "", strUrl = "";
            ArrayList<String> arrList = new ArrayList<String>();
            Pattern pattern = Pattern.compile("^(https?)
    :\\//([^:\\s]+)(:([^\\//]*))?((\\/[^\\s/\\/]+)*)?\\//([^#\\s\\?]*)(\\?([^#\\s]*))?(#(\\w*))?$");

    Matcher matcher = pattern.matcher((CharSequence) strMessage);

            if(matcher.find())
            {
                    strLink = matcher.group();
                    arrList.add(strLink);
            }

            pattern = Pattern.compile("[(http(s)?)
    :\\//\\/(www\\.)?a-zA-Z0-9@:%._\\+~-#=]{2,256}\\.[a-z]{2,6}\\b([-a-zA-Z0-9@:%_\\+.~-#?&//=]*)");

    matcher = pattern.matcher((CharSequence) strMessage);

            if(matcher.find())
            {
                    strLink = matcher.group();
                    arrList.add(strLink);
            }

            Pattern p = Pattern.compile ("((?<=src=[\"|'|`])|(?<=href=[\"|'|`]))(.+?)([\"|'|`])");
            Matcher m = p.matcher ((CharSequence) strMessage );

            while (m.find())
            {
                    strUrl = m.group ();
                    arrList.add(strUrl);

            }
            strID = message.getMessageID();

            SimpleDateFormat transFormat = new SimpleDateFormat("yyyy-MM-dd HH:mm:ss");

            strRdate = transFormat.format(message.getSentDate());

            if(strID == null) strID = "ID Empty";
            DataPro( strID, strID, strRdate, arrList);

    }

    private static void DataPro(String strId, String strSubject, String strRdate,ArrayList<String> arrList)
            throws DataServerException{
            if(arrList.size() > 0)
            {
                    dao.insertLinkData(strId, strSubject, strRdate, arrList);
            }

    }
```

Fig. 10. Link extraction module program

This program could read eml file using mail.jar in java environment, use regular expression analysis library, and use maria db 10.2.14-winx64 version for storage DB. The libraries used for this program are as follows (Fig. 9):

This program can read EML (MIME RFC 822 standard format of email programs) file using mail.jar (mail library) in java language. The significant links applied to the Regular expression were extracted from e-mail message of the EML file. Regular expression means a string formed by a constant or specific rule. It is a formal language that has a format. The extracted details are saved at Database Systems (DBMS) and the used DBMS version was the Maria DB 10.2.14-winx64.

It is a structure that, when an email body is imported, the data format is converted through a mail attribute and a mail session attribute, and a link is extracted and stored in a regular format for the link extraction pattern, as well as a structure to arrange lists after extracting the image of the text and the link included in the hyperlink and again removing duplicated links.

E-mail consists of various elements such as a simple text format, images, sound, and attachments, therefore it have to be changed to a single format and extracted links. When the links are extracted, the regular expressions are used as just described. The links are extracted from hyperlinks included in the body of the email and in-line links for visual elements and saved. It is convenient to use array to treat the extracted details in program language, but array declaration makes unit control more complicated. Therefore ArrayList that is a data structure API provided by JAVA is used. When the e-mail link is extracted, it is saved by the DB storage module as shown in Fig. 11.

```
private static void DataPro(String strId, String strSubject,ArrayList<String> arrList) throws
DataServerException{

        HashSet hs = new HashSet(arrList);
        ArrayList<String> newArrList = new ArrayList<String>(hs);

        if(hs.size() > 0)
        {
                dao.insertLinkData(strId, strSubject, arrList);
        }
}
```

Fig. 11. DB storage module

4   Link Extraction Model Verification

In order to validate the link extractor model, it is verified whether the link included in the text is successfully extracted after inputting the e-mail. A total of 136 incoming e-mails were the ones that the RBL, dynamic analysis system and vaccine blacklist of the spam system are bypassed, and 720 links were successfully extracted through the link extractor, and it is stored to check how many links of a certain mail have been

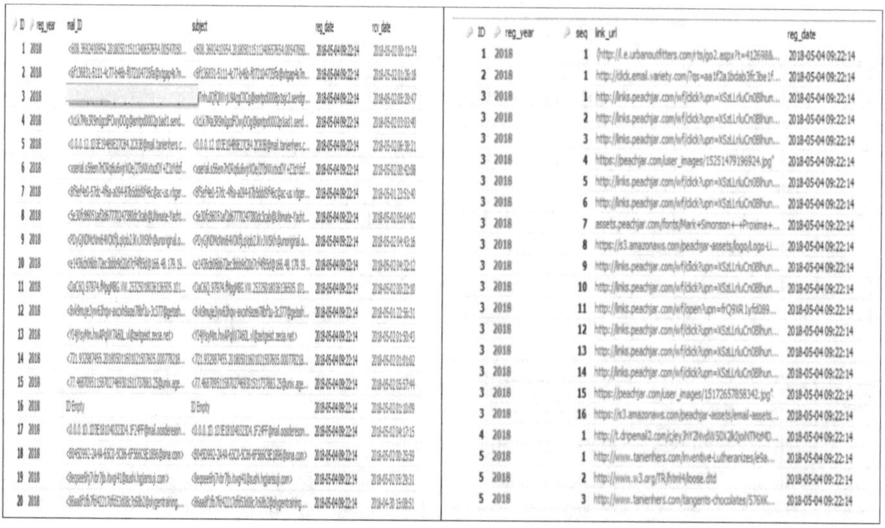

Fig. 12. Link extractor storage DB

extracted with e-mail ID imported and ID of the link extracted as shown in Fig. 12. A maximum of 71 links and a minimum of 0 links are extracted per mail, and an average of about 5 links are evicted, and most of them are less than three except for the specific mails with excessive number of links.

As a result of analyzing the 720 links stored by the security control staff, it is confirmed that there are 20 actual malicious links as shown in Table 3. It is possible to judge that the number of monitoring is not excessive except for a small number of mail including a large number of links. If security monitoring and analysis had not been carried out by storing with link extractor, it could have not been confirmed whether the whole amount was not recognized and introduced inside.

Table 3. Identified malicious links

ID	Malicious link
16	Http://****icdream.ua/CNC/scan/enx.php?email = ?young1pk@*****.kr
25	Https://****366128.wixsite.com/microsoft-update
39	Http://****bay.net.au/wisdom/index.php?email = ?*****@korea.kr
41	Https://*****lokafola.info/gehrinom/serkl.php? = id = 3D032983
55	Http://****heapbacklinks.com/dnt1/aut.php?userid = 3Dbaggage@korea.kr
57	Http://www.****festival.com/wp-admin/network/chnk/admin.php? userid=3Dwinslim@korea.kr
60	Http://****509d.ngrok.io/021/?email = vat12201570@******.go.kr
65	Http://www.****fieldcelebrant.com.au/gnhay/database/index.php
66	Https://****erweelegroup.com/wp-includes/css/cgi_bin/mail.php?main_domain=

(*continued*)

Table 3. (*continued*)

ID	Malicious link
78	Https://****dspartycompany.co.zw/upgrade/admin1/index.php? userid = fpis@*****.kr
81	Http://****boraluxuryconcierge.com/zroot (http://****boraluxuryconcierge.com/zroot)
85	Http://****wel.com/admin/superadmin/css/english%20newchn/newchn/?email = ? *****@****.com
88	Https://****anatv.com/wp-content/uploads/block
93	Http://****.labour.go.th/components/com_menus/DHL/index.php email = ? homebirds@korea.kr
101	Http://******.online/cgm1/index.php?username = ?yjung6703@korea.kr
108	Http://****oma.com/1ndex.php
119	Http://****lerstrove.com/e-banking%20update
121	Http://****myjino.ru/Korea?email = ?nierchoi@me.go.kr
133	Http://****itenergyservicesltd.com/wp-admin/johndey/index.php?email = ? ****2001@korea.kr
134	Http://www.****ayaonline.com/wp-contents/FILES/chnk/admin.php?userid=? kyb560228@korea.r

5   Conclusion and Suggestions

This study proposed a security control model that can monitor spear phishing using dynamic analysis of e-mail attached file and link bypassing security system based on blacklist matching. In conclusion, this study suggests that you devise and use a collecting plan with variously linking basic link extraction model with your company's security equipment. Also, this study wishes that you compose a security control system to register a periodic whitelist to continuously perform exception processing, and to store and manage blacklist of links determined to be harmful in order to reduce the number of evicted links. However, since the validation of this study is limited to a few samples, future research will improve the results by accumulating long-term actual operating data.

References

1. KISA: Analysis of Domestic Spear Phishing Type (2014)
2. Trend Micro Inc.: Spear-Phishing Email: Most Favored APT Attack Bait, Trend Micro Incorporated Research Paper (2012)
3. FireEye, Inc.: Spear Phishing Attacks why they are successful and how to stop them. White Paper (2012)
4. IBM, Threat Intelligence Index (2017)
5. Thomas, K., Li, F., Zand, A., Barrett, J., Ranieri, J., Invernizzi, L., Markov, Y., Comanescu, O., Eranti, V., Moscicki, A., Margolis, D.: Data Breaches, Phishing, or Malware? Understanding the Risks of Stolen Credentials, Google Research (2017)

6. Trend Micro Inc.: Spear-Phishing Email: Most Favored APT Attack Bait, Trend Micro Incorporated Research Paper (2012). Sawyer, J.H.: How Attackers Target and Exploit Social Networking Users, Dark Reading Report (2013)
7. Software Thought Leadership: Proactive response to today's advanced persistent threats, White Paper (2013)
8. Cyveillance, Malware Detection Rates for Leading AV Solutions (2010). http://www.cyveillance.com/web
9. Manning, R.: Phishing Activity Trends Report 3rd Quarter 2017. AntiPhishing Working Group (APWG) (2018)
10. Lee, M.-h., Yoo, J.-Y.: Impact of spam-mail on security and its countermeasures, Journal of Information Technology Services (2015)
11. Park, J., Cho, G.: A unknown phishing site detection method in the interior network environment. J. Korea Inst. Inf. Secur. Cryptology **25**(2), 313–320 (2015)

An Efficient Technique of Detecting Program Plagiarism Through Program Slicing

Junhyun Park[1(✉)], Hwanchul Jung[2], Jongseok Lee[3], and Jangwu Jo[4]

[1] Department of Computer Engineering, Dong-A University,
Busan 49315, Korea
wikithom91@gmail.com
[2] VODAS CO., Ltd., Seoul, Korea
karnerwels@naver.com
[3] Department of Computer Engineering,
Woosuk University, Wanju, Jeollabuk-do 55338, Korea
jong1007@woosuk.ac.kr
[4] Department of Computer Engineering,
Dong-A University, Busan 49315, Korea
jwjo@dau.ac.kr

Abstract. In this paper, we survey state of the art in plagiarism detection techniques, called GPLAG. This technique can detect five popular plagiarism techniques, such as format alteration, identifier renaming, statement reordering, control replacement, and code insertion. The problem of this technique is that it becomes inefficient when PDG grows. To resolve a problem of time complexity, this paper proposes a way to perform program slicing first and PDG comparison later. Original program and plagiarized program are sliced respectively, then PDG of original program's slice is compared with PDG of plagiarized program's slice. Since program slicing reduces the size of PDG, time to compare PDGs can be reduced. We choose most used variables as slicing criterion, and we can maintain accuracy even after program slicing. By experiments we show that efficiency is enhanced and accuracy is also maintained.

1 Introduction

The current plagiarism detection techniques can be classified into two kinds: syntax based and semantic based. String comparison, AST(Abstract Syntax Tree) comparison, and token comparison are examples of syntax based method, and PDG(Program Dependence Graph) comparison is an example of semantic based one [1–6]. Syntax based one is weak to plagiarism techniques such as statements reordering, control replacement, and code insertion. Semantic based method can handle the above plagiarism techniques [4].

The problem of the semantic based one is that it becomes inefficient when PDG grows. This technique compares PDGs to find the graph isomorphism or

© Springer Nature Switzerland AG 2019
R. Lee (Ed.): SNPD 2018, SCI 790, pp. 164–175, 2019.
https://doi.org/10.1007/978-3-319-98367-7_13

subgraph isomorphism. The time complexity of graph isomorphism is known to be NP-complete. There have been several researches to reduce the number of graph comparison to improve speed [4,7]. To enhance time complexity, this paper proposes a way to perform program slicing first and PDG comparison later. Original program and plagiarized program are sliced respectively, then PDG of original program's slice is compared with PDG of plagiarized program's slice. Since program slicing reduces the size of PDG, time to compare PDGs can be reduced. We choose most used variables as slicing criterion, and we can maintain accuracy even after program slicing. By experiments we show that efficiency is enhanced and accuracy is also maintained.

The remainer of this paper is organized as follows. Section 2 reviews state of the art in plagiarism detection techniques. We suggest our plagiarism detection techniques in Sect. 3. Section 4 explains our proposed method by examples. The experimental results are presented in Sect. 5. Finally, Sect. 6 summarizes the conclusions of this study.

2   Detection of Program Plagiarism by Program Dependence Graph

Figure 1 shows the process of PDG(Program Dependence Graph) based plagiarism detection [4]. Given as input an original program P and a plagiarism program P', it generates a pair of a set of PDGs, N and N'. Then it performs

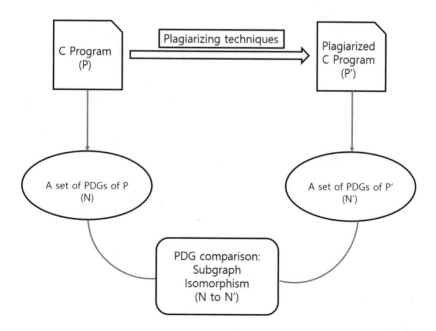

Fig. 1. PDG based plagiarism detection

subgraph isomorphism testing. Because subgraph isomorphism is NP-complete in general, it is shown to be computationally intractable. [4] tried to solve this computational intractability by introducing two level filters.

3   Plagiarism Detection Through Program Slice

This section suggests our method that improves efficiency of PDG based program plagiarism detection, even though our method keeps accuracy. Figure 2 depicts the process that adds program slicing to current PDG based program detection technique. Just like Fig. 1, it takes as input original program P and plagiarized program P'. We compute SP that is a slice of program P, and SP' that is a slice of plagiarized program P'. We choose slicing criterion among return values or parameters, because it's hard to change return values or parameters in plagiarized program. The rest of process is same as Fig. 1. Compared with Fig. 1, our method decreases the number of nodes and edges of PDG. Reducing the size of PDG increases the efficiency of detecting subgraph isomorphism. Next section shows the example of reducing the size of PDG.

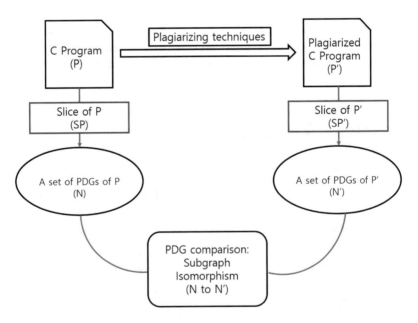

Fig. 2. Program slice to reduce the size of PDG

4   Example

Table 1 shows original program P and Table 2 shows its plagiarized program P'. Program P' is a plagiarized version of original program P by statement

Table 1. Original program P

```
int  write(int  a)  {
    return  a;
}
main()  {
    int  i;
    int  n  =  10;
    int  sum  =  1;
    int  product  =  1;

    i  =  0;
    while(i  <  n)  {
        product  =  product  +  i;
        sum  =  sum  +  i;
        i++;
    }

    write  (sum);
    write  (product);

    return  sum;
}
```

Table 2. Plagiarized program P'

```
int  write(int  a)  {
    return  a;
}
main()  {
    int  um  =  1;
    int  i;
    int  roduct  =  1;
    int  temp;
    int  n  =  10;

    i  =  0;
    while(i  <  n)  {
        um  +=   i;
        roduct  =  roduct  +  i;
        temp  =  um;
        temp  =  product  +  temp;
        i++;
    }
    write  (roduct);
    write  (um);
    return  um;
}
```

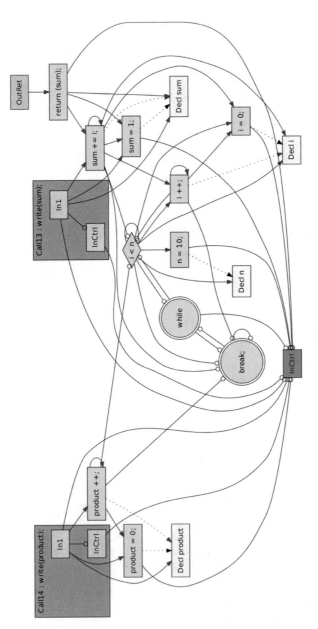

Fig. 3. PDG of original program P

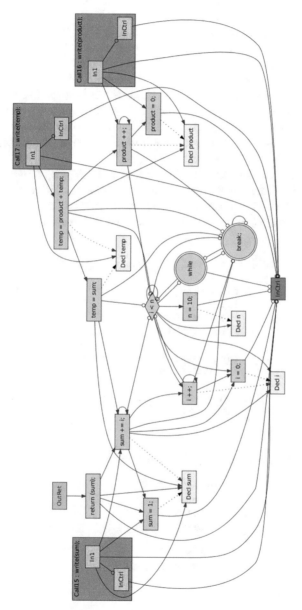

Fig. 4. PDG of plagiarized program *P'*

reordering, identifier renaming, control replacement, and code insertion. Code insertion makes plagiarized program P' larger than original program P. It causes PDG of plagiarized program to grow bigger.

Figure 3 shows PDG of original program P and Fig. 4 shows PDG of plagiarized program P'. Comparing nodes and edges in Figs. 3 and 4, two PDGs are isomorphically homomorphic except for the part of PDG from inserted code.

A program slice consists of the parts of a program that (potentially) affect the values computed at some point of interest, referred to as a slicing criterion [8]. Table 3 shows a slice of original program P in Table 1 with respect to criterion (18, sum). As can be seen in the Table 3, all statements involving variable **sum** have been sliced away. Table 4 also shows a slice of plagiarized program P' in Table 2 with respect to criterion (17, sum). We choose return value as slicing criterion. One of the most interesting thing is that the slice of plagiarized program SP' does not contain codes inserted in P'.

Figure 5 shows PDG of Table 3. The size of PDG in Fig. 4 become smaller than the size of PDG in Fig. 3. The reduction of size of PDG causes subgraph isomorphic detection to be efficient. Figure 6 shows PDG of Table 4. The PDG of Table 4 is similar to that of Table 3, except for the difference of variable names.

Table 3. SP, a slice of original program P

```
int  main () {
    int  i ;
    int  n =  10;
    int  sum =  1;

    for ( i =  0; i < n; i++) {
        sum = sum + i ;
    }
    write  (sum );
    return  sum;
}
```

Table 4. SP', a slice of plagiarized program P'

```
int  main () {
    int  um =  1;
    int  i ;
    int  n =  10;

    i =  0;
    while ( i < n ) {
        um +=   i ;
        i++;
    }
    write  (um);
    return  um;
}
```

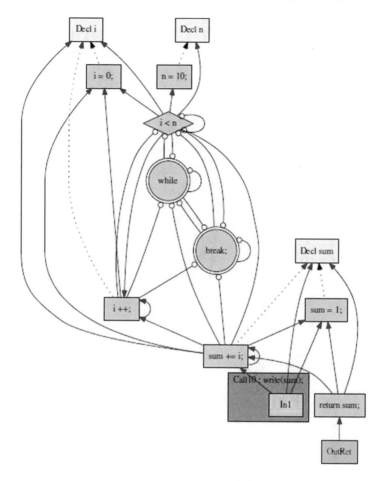

Fig. 5. PDG of Table 3

5   Experiments

5.1   Setup for Experiments

We used *Frama-C* for experiments. *Frama-C* is a suite of tools dedicated to the analysis of the source code of software written in C [9]. *Frama-C* provides libraries for slicing programs, generating PDGs, and performing subgraph isomorphism.

5.2   Programs for Experiments

We chose C programs that are common commands in Linux, whose characteristics are listed in Table 5. The number of functions which exclude main function and usage function is 6 and 5 in cp.c and mv.c, respectively.

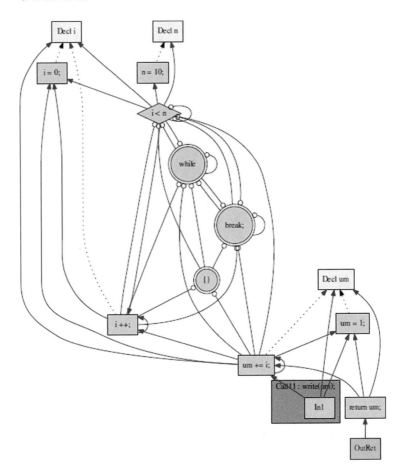

Fig. 6. PDG of Table 4

5.3   Plagiarized Programs

We manually plagiarized programs in Table 5. We use common plagiarism techniques: statement reordering, control replacement, code insertion. Table 6 shows
the number of plagiarism techniques applied to each function. Code insertion
is not applied to the following functions: cp_option_init, re_protect, and tar-
get_directory_operand. Function named target_directory_operand is not changed.

5.4   PDG Comparison

We generate PDGs of both original programs and plagiarized ones. Table 7 compares PDG of original programs with PDG of plagiarism programs in terms of
the number of nodes and edges. Table 7 shows that code insertion introduces
new nodes and edges.

Table 5. Programs for experiments

Program	Line of code	# of function	Description
cp.c	1175	6	Copy files
mv.c	497	5	Move files

Table 6. Plagiarism techniques applied to programs

Name of function	Plaglarism techniques		
	Statement ordering	Control replacement	Code insertion
cp_option_init	10	0	0
decode_preserve_arg	7	1	3
do_copy	1	5	6
make_dir_parents_private	7	6	4
re_protect	1	8	0
target_directory_operand	0	0	0

Table 7. PDG comparison before slicing

Name of function	Original		Plagizrized	
	Node	Edge	Node	Edge
cp_option_init	44	111	44	111
decode_preserve_arg	99	313	106	345
do_copy	417	2117	434	2171
make_dir_parents_private	373	1588	387	1635
re_protect	213	675	213	675
target_directory_operand	109	211	109	211

Table 8. PDG comparison after slicing

Name of function	Slice of original		Slice of plagizrized	
	Node	Edge	Node	Edge
cp_option_init	40	104	40	104
decode_preserve_arg	88	290	88	290
do_copy	163	735	163	735
make_dir_parents_private	126	338	126	338
re_protect	51	101	51	101
target_directory_operand	25	41	25	41

We also generate PDGs of slices of original and plagiarized programs. Table 8 shows the number of nodes and edges of PDGs of two slices. And the number of nodes and edges of original programs are same as that of plagiarized one. Therefore, the PDG comparison after slicing is efficient.

5.5   Accuracy

Table 9 compares the accuracy of the existing PDG-based detection method with the accuracy of proposed PDG-based detection method. Current one can find graph isomorphism between original programs and plagiarized ones. Our method can keep the accuracy because it can find the graph isomorphism between slice of original program and slice of plagiarized program.

Table 9. Comparison of accuracy

Name of function	Subgraph isomorphism	
	Current one	Our one
cp_option_init	O	O
decode_preserve_arg	O	O
do_copy	O	O
make_dir_parents_private	O	O
re_protect	O	O
target_directory_operand	O	O

5.6   Efficiency

Table 10 compares the two analyses in terms of analysis time. The analysis time consists of finding subgraph isomorphism. Table 10 shows that our approach is up to 8 times faster than existing PDG based approach. This is due to the fact that PDG of sliced programs is smaller than that of original ones.

Table 10. Comparison of efficiency

Program	Current one	Our one
cp.c	7(s)	1(s)
mv.c	4(s)	0.5(s)

6   Conclusion

In this paper, we surveyed state of the art in plagiarism detection techniques, called GPLAG. This technique can detect five popular plagiarism techniques, such as format alteration, identifier renaming, statement reordering, control replacement, and code insertion. This paper suggested a method that performs program slicing before generating PDG in [7], in order to enhance the efficiency of detecting subgraph isomorphism. In experiments, we can see our method is 8 times faster than that of [7]. This experiments also show that our method has same accuracy as [7], although the program slicing makes PDG small.

Acknowledgement. This work was supported by the Korea Institute of Energy Technology Evaluation and Planning (KETEP) and the Ministry of Trade, Industry Energy (MOTIE) of the Republic of Korea 20161210200500.

References

1. Baker, B.S.: On finding duplication and near duplication in large software system. In: Proceedings of 2nd Working Conference on Reverse Engineering (1995)
2. Faidhi, J.A.W., Robinson, S.K.: An empirical approach for detecting program similarity and plagiarism within a university programming environment. Comput. Educ. **11**, 11–19 (1987)
3. Roy, C.K., Cordy, J.R.: A Survey on Software Clone Detection Research. School of Computing Queen's university at Kingston Qntario, No. 2007–541 (2007)
4. Liu, C., Chen, C., Han, J.: GPLAG: detection of software plagiarism by program dependence graph analysis. In: Proceedings of the 12th ACM SIGKDD International Conference on Knowledge Discovery and Data Mining (2006)
5. Parker, A., Hamblen, J.: Computer algorithms for plagiarism detection. IEEE Trans. **32**, 94–99 (1989)
6. Prechelt, L., Malpohl, G., Philippsen, M.: Finding plagiarism among a set if programs with JPlag. J. Univ. Comput. Sci. **8**(11) (2002)
7. Kim, S., Han, T.: Plagiarism detection using dependency graph analysis specialized for javascript. J. Korea Soc. Comput. Inf. **37**(5), 394–402 (2010)
8. Tip, F.: A Survey of Program Slicing Techniques. CWI (Centre for Mathematics and Computer Science) Amsterdam (1994)
9. Frama-C user manual. https://frama-c.com/download/frama-c-user-manual.pdf

Author Index

© Springer Nature Switzerland AG 2019
R. Lee (Ed.): SNPD 2018, SCI 790, p. 177, 2019.
https://doi.org/10.1007/978-3-319-98367-7

Printed in the United States
By Bookmasters